INTERNATIONAL ASSIGNMENTS
An Integration of Strategy, Research, and Practice

Linda K. Stroh
Loyola University Chicago

J. Stewart Black
Center for Global Assignments

Mark E. Mendenhall
The University of Tennessee at Chattanooga

Hal B. Gregersen
Brigham Young University

LAWRENCE ERLBAUM ASSOCIATES, PUBLISHERS
2005   Mahwah, New Jersey   London

Copyright © 2005 by Lawrence Erlbaum Associates, Inc.
All rights reserved. No part of this book may be reproduced in any form, by photostat, microform, retrieval system, or any other means, without the prior written permission of the publisher.

Lawrence Erlbaum Associates, Inc., Publishers
10 Industrial Avenue
Mahwah, New Jersey 07430

Cover design by Sean Trane Sciarrone

Library of Congress Cataloging-in-Publication Data

International assignments : an integration of strategy, research, and practice / Linda K. Stroh ... [et al.].
   p. cm.
Includes bibliographical references and index.
ISBN 0-8058-5049-X (alk. paper) — ISBN 0-8058-5050-3 (alk. paper)
  1. Executives—Training of. 2. Leadership—Study and teaching. 3. International business enterprises—Employees—Training. 4. International business enterprises—Personnel management. 5. Cross-cultural orientation. I. Stroh, Linda K.

HD30.4.I574 2004
658.4'07—dc22
                                    2004050610
                                          CIP

Books published by Lawrence Erlbaum Associates are printed on acid-free paper, and their bindings are chosen for strength and durability.

Printed in the United States of America
10 9 8 7 6 5 4 3 2 1

Contents

Foreword	vii
Preface	xi
About the Authors	xv
I INTRODUCTION	**1**
1 The Strategic Roles of International Assignments in Globalization	3
2 The Process of Cross-Cultural Adjustment	29
II BEFORE THE ASSIGNMENT	**47**
3 Selecting: Identifying Candidates With Global Leadership Potential	49
4 Training: Helping People Learn to Do the Right Things	80

III DURING THE ASSIGNMENT — 97

5 Adjusting: Developing New Mental Road Maps and Cultural Skills — 99

6 Integrating: Balancing Multiple Allegiances — 120

7 Appraising: Determining if People Are Doing the Right Things — 144

8 Rewarding: Compensation and Pay — 171

IV AFTER THE ASSIGNMENT — 187

9 Repatriating: Helping People Readjust and Perform — 189

10 Retaining: Utilizing the Experienced Global Manager — 218

11 Managing the Entire Global Assignment Cycle: Establishing Best Practices — 235

References — 257

Author Index — 265

Subject Index — 269

Foreword

Although the authors of this book and I approach the strategic challenges of globalization from different perspectives, we've come to a similar conclusion: that the "people" imperative is often overlooked when companies decide to enter new countries or world regions. Stroh, Black, Mendenhall, and Gregersen base their assertion on extensive research and practice; mine is the product of more than 30 years in corporate leadership roles—the last two decades of which have included positions with substantial exposure to the global marketplace.

It is stunning to think just how much the world of commerce has changed during that time. Many of our most cherished business paradigms have been forever altered by the impact of forces such as improved communications and transportation technologies and borderless trade policies. When we speak today of the "Big Three" automakers in the United States, for example, we must remember that one of those—DaimlerChrysler—is now controlled by German ownership; and we must recognize that in any given month, a Japanese firm with U.S. manufacturing facilities is likely to be one of the top three.

As noted by Stroh et al., globalization has become a fact of business life. It is inevitable in an era when the corporate customers that many companies serve are positioning themselves to become global players. A company can no longer afford to focus on a single geographic market—not if it expects to earn business contracts from corporations and conglomerates that are aggressively developing their business opportunities on a global stage. Such customers won't wait for their traditional suppliers to catch up. Instead, they will turn to new suppliers, suppliers that can in fact help produce and distribute their products and services in the markets they want to enter.

Nor are customer demands the only significant force driving the trend toward globalization today. Foreign competitors play a key role, too. As industries consolidate in Europe, North America, and Asia, we have seen time and again how the largest players in each of those markets begin to look offshore for their next big growth opportunities. Precisely because these competitors have developed world-class scale and processes in their home markets, they can represent a significant threat to companies in other world regions. This is especially true when local companies fail to recognize the threat. If local companies do not move aggressively to become world-class competitors themselves, their domestic markets can evaporate in a heartbeat.

So perhaps the first important lesson of the past 20 years is that global presence has become a strategic imperative for major companies today. It is not merely a "nice-to-have" option for stimulating growth. It is a "must-have" capability for ensuring the company's long-term survival.

Both the authors of this book and I know this lesson has important implications for a company's leadership. The reason is simple: Leadership development is—and always has been—one of the chief executive officer's (CEO's) most important responsibilities. Preparing the next generation of company leaders is every bit as critical as producing great products or developing an effective strategic plan when it comes to ensuring a company's future competitiveness. Today, that means preparing leaders who can understand and operate effectively in a global business environment.

All too often, however, this responsibility is overlooked. A survey of business literature over the past couple of decades will reveal no shortage of material about strategies for globalization: how to globalize products, how to build a global supply chain, how to manage projects across continents and time zones, and so on. However, comparatively little information exists on the subject of developing a management group that can effectively accomplish these things. I believe that is because few business experts understand how to do it—or even recognize that it ought to be a priority.

It is hard work to develop a cadre of globally competent leaders and managers. It takes planning and commitment. In fact, as noted in this book, it has to begin very early in the careers of fast-track managers so that they have the chance to develop a broad base of cross-cultural experience and insight before they are called to take on vital leadership roles in overseas operations. The CEO must set the stage for all this to happen by insisting that the company's leadership development process includes a component for globalizing executives. Nothing short of such a commitment will produce the pool of candidates necessary for the company to succeed as a truly global enterprise.

In fact, I believe that companies ought to do even more than develop effective strategies for globalizing their next generation of leaders. Tremendous benefits can be gained by taking the process a step further—by globalizing the company's entire culture so that its vision, values, and standards are understood and practiced in every one of its markets everywhere in the world. Once again, the insight gained from my experience is consistent with the content of this Stroh et al. book.

That is the second important lesson to be drawn from the past 20 years. Companies that hope to succeed in other countries and regions must make a global commitment to their vision, values, and standards. In case after case, major companies have learned that it is impossible to run a successful business in a foreign market by remote control. They cannot compromise on the quality of the people they send there or on the people they hire there. They cannot compromise on the thoroughness of their planning, either—or on the resources they are willing to commit to the development of the market.

This is not to say the solution is to transplant a set of home-country business rules in overseas operations. On the contrary; consistent with this book's perspective, global companies must demonstrate a deep and profound respect for the customs, laws, and dynamics of each market in which they operate. However, what they can transplant—what they must transplant—is a clear sense of their corporate vision, values, and standards of excellence.

It is absolutely essential that a company's leaders establish this common ground for employees in every market around the world. Employees have to clearly understand the global strategy and their role in executing that strategy. Just as important, they have to understand how they are expected to behave as members of a global team—how people are to be treated and how the work gets done. When values such as these are developed and shared across the enterprise, it has an energizing effect on the entire organization. In a very real sense, a strong corporate culture can help to offset the distance between and among far-flung operations around the world. It can accelerate the company's success, too—by ensuring that the employees of every work team and every business unit are pulling in the same direction.

What we come to, then, is a full appreciation of the people imperative identified in this book. Experience shows that any globalization initiative is destined to produce disappointing results unless it includes a strong focus on developing effective global managers and leaders. That is what makes the content of this book critically important: It identifies leadership that every executive would do well to heed.

—William L. Davis
President, Chairman and Chief Executive Officer
RR Donnelley
November, 2003

Preface

That rapid globalization is taking place in businesses around the world is beyond dispute. If multinational firms are to prosper now and in the future, their managers must be able to function in a global context—formulating and implementing strategies, inventing and utilizing technologies, and creating and coordinating information. International assignments are the single most powerful means for developing future global leaders.

Although increasing numbers of corporations recognize the value of developing global leaders, all too often the business objectives of international assignments are not being met. In short, firms are not getting good returns on the time and expense they invest in international assignments. Uncompleted assignments, poor performance, and high rates of turnover after repatriation are common and cost even moderately sized multinational corporations tens of millions of dollars each year. Additionally, individuals are not getting good returns on their investment in accepting international assignments. Seven out of 10 managers believe that their time spent on an international assignment had a negative impact on their career. The reality is that international assignments are stressful and that stress shows up in substance abuse, strained and broken marriages, and other family problems.

In addition to studying international assignments, we, as the authors of this book, all have experienced and seen firsthand the significant losses and profits to organizations and to individuals that result when international assignments are poorly or well designed. These experiences have given us an understanding of the issues as well as an intense motivation to provide a systematic guide for constructing effective international assignment systems—systems that achieve critical competitive results today and develop the global leaders of tomorrow.

Although firms have been sending employees on international assignments for decades, systematic understanding is sorely lacking in such critical aspects of the assignment process as the selection process; the training required; factors that affect adjustment, performance, and commitment; and how to retain and capitalize on international experience once employees return home. Although there has been increased scholarly research on international assignments, much of it is tucked away in academic journals and is unknown to the very executives who would most practically benefit from that knowledge. We have written this book with the hope that by learning the best practices and latest research findings on international assignments, both the individuals embarking on international assignments and the organizations sending them will experience more positive results.

This book was written from a North American multinational corporation focus; however, many examples are used from Finnish, Japanese, Brazilian, Malaysian, and other European, Asian, and Southeast Asian experiences and research. This book would be most appropriate for managers working in multinational firms who deal with international assignees. This book would also be excellent for advanced undergraduate and graduate students in MBA, Human Resources, Organizational Development, Industrial Organizational Psychology, Organizational Behavior, or Cross-Cultural studies programs. This book accesses much information that was used in an earlier book by the same authors. The title of that book is *Globalizing People Through International Assignments* (1999).

AUDIENCE

This book is written primarily for executives whose focus includes the global economy and the strategic role of people in achieving international competitiveness. This includes chief executive officers, line managers, and especially human resource management executives. After reading this book, executives should have a sophisticated but practical understanding of the strategic roles of international assignments as well as of the complete cycle of these assignments.

Each chapter has four objectives: (a) to examine a specific problem concerning international assignments, (b) to explain the underlying principles for understanding the problem, (c) to provide a framework for analyzing the issue, and (d) to present recommendations for executives to follow in redesigning or enhancing their current systems.

It will also serve as a textbook for courses on International Human Resources.

OVERVIEW OF CONTENTS

The book covers every major aspect or stage of international assignments. Chapter 1 defines the strategic roles that international assignments can play and discusses how these roles need to be adjusted, depending on the firm's particular stage of internationalization. Chapter 2 explores the process of cross-cultural adjustment, critical for effectively working and living in foreign cultures. Chapter 3 examines issues pertaining to selecting people for international assignments, addressing both who should be selected and how they should be chosen. Chapter 4 focuses on the challenge of training people so that they perform effectively while overseas. Specifically, the chapter provides a framework for firms so that they do not over- or underinvest in these training needs. Chapter 5 focuses on the factors that affect successful cross-cultural adjustment and describes ways firms can facilitate the adjustment process. Chapter 6 provides pioneering insights into the dynamics of dual allegiance (to the parent firm and the local foreign operation) that occurs among international managers and discusses how firms can foster "dual citizens," or employees with dual allegiance to both their home-country operation and local unit. Chapter 7 examines the difficulties of appraising employees while they are on foreign assignments and provides a model of how this can be done effectively. Chapter 8 explores the problems most firms face regarding the high compensation costs of maintaining employees who work abroad and outlines a means of significantly reducing those costs while motivating employees to accept and perform well in international assignments. Chapter 9 unravels the process of coming home, exploring the factors that affect repatriation adjustment and job performance. It points to specific steps firms can take to facilitate repatriation adjustment and performance. Chapter 10 explains the factors that determine whether high-performing repatriated managers will stay with the firm or look for greener pastures elsewhere. Included are recommendations for retaining and utilizing high-performing returning managers and executives. Finally, chapter 11 summarizes and integrates critical elements of the international assignment cycle. It also describes firms that represent some of the best practices in selection, training, appraisal, compensation, and repatriation, as well as security issues companies must address now more than ever.

ACKNOWLEDGMENTS

There are many people to whom we owe thanks. First, we thank our families for their support and understanding. We thank them for their willingness to live in such places as Japan, Finland, Switzerland, and Canada. We also thank them for making international travel a possibility, as they sometimes travel with us, and for maintaining the home front when we are apart.

We are also thankful to the Fortune® 500 executives who have written the "From the Front Line" boxes for this book. In particular, we thank Raj Tatta and John Neylon of PriceWaterhouseCoopers, an anonymous contributor from a financial services company, Maureen Ausura from ADM, Jim Pilarski from Marriott International, Kevin Gazarra from Intel, Jenny Li from Sony Pictures Entertainment, Dick Bahner from R Bahner International, Kerry Weinger from

Baker and McKenzie, Marie Howard from The Procter & Gamble Company, John Murphy from Motorola, Amy Glynn from Dow Jones & Company, and Shlomo Ben-Hur and Kerstin Boecker from DaimlerChrysler. Their time and contributions have set this book apart from others.

We are also grateful to colleagues at Loyola University Chicago; University of Tennessee, Chattanooga; The University of Michigan; and Brigham Young University. In particular, we thank Stacey Bogumil, Kelly Chew, Antoinette Cromartie, and Brooke Zahara who collected company data for chapter 3 (this volume) as part of a group research project they did for a master's level class in the Human Resources & Industrial Relations program (HRIR 443) at Loyola University Chicago.

In addition, we thank Anne Duffy, our Lawrence Erlbaum Associates, Inc. project director, for her constant support and encouragement. It was a real treat working with Anne. Finally we thank Erica Fox for her expert research and editing skills. This book could not have happened without Erica's major contribution.

Linda K. Stroh
Loyola University Chicago

J. Stewart Black
Center for Global Assignments

Mark E. Mendenhall
University of Tennessee, Chattanooga

Hal B. Gregersen
Brigham Young University

About the Authors

Linda K. Stroh is a Loyola University Faculty Scholar and Professor of Human Resources & Industrial Relations at the Graduate School of Business, Loyola University Chicago. Stroh received her PhD from Northwestern University in Human Development. She also has a Postdoctorate in Organization Behavior from Northwestern's Kellogg Graduate School of Management. She received her BA from McGill University in Montreal, Quebec, Canada and her MA from Concordia University also in Montreal.

Stroh has taught and published over 100 articles, books, and technical reports on issues related to domestic and international organizational behavior issues. Linda's work can be found in journals such as *Strategic Management Journal, Journal of Applied Psychology, Personnel Psychology, Academy of Management Journal, Journal of Vocational Behavior, International Journal of Human Resource Management, Journal of Organizational Behavior, Journal of Management Education, Sloan Management Review, Human Resource Management Journal, Journal of World Business,* and various other journals. In addition to this book, Stroh is coauthor of two other books, *Globalizing People Through International Assignments* and *Organizational Behavior: A Management Challenge.*

Stroh was honored at the 2000 Academy of Management Meeting with the Sage publications research scholar award. She was also named the Graduate Faculty Member of the Year at Loyola University Chicago (2000) and was selected as a Loyola University Faculty Scholar in 2001. *The Wall Street Journal, The New York Times, The Washington Post, The Chicago Tribune, Fortune, Newsweek, U.S. News and World Report,* and *Business Week* as well as various other news and popular press outlets have cited Stroh's work. Stroh's research has also been featured several times on *NBC Nightly News with Tom Brokaw* and CNN.

In addition to her teaching and research, Stroh serves as the Academic Adviser for the International Personnel Association (an association of 60 of the top 100 multinational companies in the United States and Canada). She is also a past Chair of the Careers Division for the Academy of Management. Linda currently serves on the editorial review board for the *Journal of Applied Psychology, Journal of World Business, Journal of Vocational Behavior,* and *Organizational Analysis.* She also served on SHRM's International Human Resources Task Force.

Stroh has consulted with over 40 *Fortune* 500 organizations regarding such issues as motivation, leadership, change management, problem solving, strategic planning, diversity, international assignments, and cross-cultural management.

Linda and her husband, Greg, have married children: Angie and Joe Gittleman and Brad and Brandy Stroh. They are awaiting their first grandchild in July.

J. Stewart Black is the Managing Director of the Global Leadership Institute. He is also Executive Director of the Asia Pacific Human Resource Partnership for the University of Michigan and a Professor of Business Administration for the Business School.

Black received his undergraduate degree in Psychology and English from Brigham Young University where he graduated with honors. He earned his master's degree from the Business School at Brigham Young University where he was on the dean's list and graduated with distinction. After graduation, he worked for a Japanese consulting firm where he eventually held the position of managing director. Black returned to the United States and received his PhD from the University of California, Irvine. He then took a position as assistant professor at the Amos Tuck School of Business Administration, Dartmouth College and was later promoted to associate professor. After 5 years at Dartmouth College, Black accepted a position as associate professor of international management at Thunderbird (The American Graduate School of International Management). Black also served as the director of the Thunderbird Japan Campus and later as the Associate Vice President for Thunderbird Executive Education. Prior to leaving Thunderbird to join the University of Michigan, Black was promoted to full professor with tenure.

Black is a leading instructor and scholar in strategic change, globalization, leadership, and international human resource management. His research and consulting focuses on the areas of change, global leadership, strategic human resource management, international assignments, and cross-cultural management.

Black has traveled extensively throughout Europe and Asia. In addition, he has lived and worked in Japan for over 4 years and speaks Japanese fluently. He has been a visiting professor at International University of Japan on three different occasions.

Black has consulted with and done seminars for a variety of international firms in the areas of global leadership and international human resource management including American Express, Black & Decker, The Boeing Company, Brunswick,

Dofasco, ExxonMobil, General Motors, IBM, The Kellogg Company, Motorola, NASA, Solar Turbines, TRW, Honda Motors, Isuzu Motors, Kawasaki Shipping, Kawasaki Steel, Nissan Motors, Nihon Unisys, Sunkyong, and TDK.

Black is also a coauthor of eight books including *Leading Strategic Change: Breaking Through the Brain Barrier; Global Explorers: The Next Generation of Leader; Globalizing People Through International Assignments; So You're Going Overseas: A Handbook for Personal and Professional Success; Management: Meeting New Challenges;* and *International Business Environments.* He is the author of over 50 articles in the area of international human resource management that have appeared in both managerial and academic publications: *Business Week, The Wall Street Journal, Fortune, Workforce, International Business, Mobility, Personnel, Academy of Management Review, Academy of Management Journal, Human Resource Management, Group & Organization Studies, International Journal of Intercultural Relations, Asia-Pacific Journal of Management, Journal of International Business Studies, Academy of Management Journal,* and *Human Relations.* He has also made over 40 presentations at professional meetings in the United States and throughout the world.

He is a member of the Academy of Management and has served on the Executive Committee of the International Management Division. He has served as Editor of the *Journal of International Management* and an Editorial Board Member of the *Academy of Management Review.* He has also served as a reviewer for *Administrative Science Quarterly, Journal of International Business Studies,* and *Academy of Management Journal.*

Mark E. Mendenhall holds the J. Burton Frierson Chair of Excellence in Business Leadership at the University of Tennessee, Chattanooga. He received his BS degree (1980) in psychology and his PhD degree (1983) in social psychology, both from Brigham Young University. In 1998, he held the Ludwig Erhard Stiftungsprofessur Endowed Chair at the University of Bayreuth (Germany), and since 1999, he has been a visiting professor on the faculty of the Europa Institute at the University of Saarland (Germany).

Mendenhall is an internationally recognized scholar in the field of international human resource management, his areas of expertise being the cross-cultural adjustment of expatriate managers and global leadership development. His other research and consulting interests are in the areas of leadership and organizational change and the nonlinear dynamics of organizational systems.

He has authored the following books: (with G. Stahl) *Managing Culture and Human Resources in Mergers and Acquisitions* (forthcoming, Stanford University Press); *Blackwell Handbook of Global Management: A Guide to Managing Complexity* (with H. Lane, M. Maznevski, & J. McNett, Blackwell); *Developing Global Business Leaders: Policies, Processes, and Innovations* (2001, with T. Kühlmann and G. Stahl; Quorum); *Readings and Cases in International Human Resource Management* (2000, 3rd ed., South-Western), *Developing People*

Through International Assignments (1999, with J. Stewart Black, Hal Gregersen, and Linda Stroh; Addison-Wesley); *Cases in International Organizational Behavior* (1999, with Gary Oddou, Blackwell); *Global Management* (1995, Blackwell; with D. Ricks and B. J. Punnett); *Global Assignments: Successfully Expatriating and Repatriating International Managers* (1992, Jossey-Bass; with J. Stewart Black and Hal Gregersen); and *Readings and Cases in International Human Resource Management* (1991, 2nd ed., with G. Oddou, South-Western).

He has published numerous research articles that have appeared in such journals as *Sloan Management Review, Academy of Management Review, Journal of International Business Studies, Organizational Dynamics, Management International Review, Columbia Journal of World Business, Human Relations, Organizational Dynamics, Journal of Management Inquiry, Journal of Management History, Human Resource Management Review, Handbook of Intercultural Training, American Behavioral Scientist, Human Resource Management, Group and Organization Studies, International Journal of Management, Business Horizons, Human Resource Development Quarterly, Journal of Management Education, Asia-Pacific Journal of Management, Training and Development Journal, International Journal of Intercultural Relations, Journal of Social Psychology, Journal of European Industrial Training,* and *Teaching of Psychology.*

He is active in the Academy of Management and is currently past president of the International Division of that organization. Mendenhall actively consults and trains in the areas of global leadership and cross-cultural management. Some of the companies he has worked with include IBM-Asia Pacific, IBM-Japan, National Aeronautic and Space Administration (NASA), Boeing, Monsanto, J.C. Bamford (JCB), The Dixie Group, and Japan Air Lines Corporate Academy. Mendenhall has traveled widely and has lived overseas for 9 years (New Zealand, Japan, Switzerland, and Germany). He and his wife Janet have four children—Anthony, Nicole, Alexis, and Zachary. Anthony is currently serving a 2-year missionary stint in the West Indies and currently is living and working in Guyana.

Hal B. Gregersen is the Donald L. Staheli Professor of Global Leadership in the Marriott School at Brigham Young University. Gregersen taught previously in the Amos Tuck School of Business Administration at Dartmouth College, Pennsylvania State University, Thunderbird, Helsinki School of Economics, and completed a Fulbright Fellowship at the Turku School of Economics in Finland. He also teaches in executive education programs for numerous universities (University of Michigan, Helsinki School of Economics, the Fletcher School of Law and Diplomacy at Tufts University, and Thunderbird) and companies (Black & Decker, Cemex, EDS, IBM, Intel, Lockheed Martin, Marriott International, and Sun Microsystems). Gregersen received a PhD in Business Administration from the University of California, Irvine. In 1996, he received the Ascendant Scholar Award from the Western Academy of Management. Gregersen is also the recipient of several outstanding teacher awards.

Gregersen's primary research, teaching, and consulting interests focus on implementing global strategies and developing executive leadership capability. He is the author of seven books and more than 50 articles, book chapters, and cases on these topics and has published in top business journals such as *Harvard Business Review, Sloan Management Review, Academy of Management Journal, Journal of International Business Studies, Journal of International Management, Journal of Applied Psychology,* and *Journal of Management.*

Gregersen recently completed a major international research project for a new book, *Leading Strategic Change: Breaking Through the Brain Barrier* (with J. S. Black, Financial Times/Prentice Hall, 2003). Gregersen is also the coauthor of *Global Explorers: The Next Generation of Leaders* (with J. S. Black & A. Morrison, Routledge, 1999), a book based on in-depth interviews of exemplar global leaders and how leading multinationals develop them. In addition, Gregersen coauthored the *So You're Going Overseas* and *So You're Coming Home* books (with J. S. Black, Global Business), a series of practical books to successfully guide employees and their families through the entire international assignment cycle. Gregersen has appeared on *CNNfn* to discuss his research on global leadership and has been cited in numerous publications such as *Across the Board, Business Week, Executive Excellence, Fortune, Industry Week, Investor's Daily, Los Angeles Times, U.S. News and World Report,* and *The Wall Street Journal.*

Gregersen consults with a variety of North American, European, and Asian firms (for example, Ameritech, Intel, LG Group, Marriott International, and Nokia) to formulate and implement successful global strategies. He regularly delivers keynote speeches on international topics for firms and professional associations around the world.

Gregersen and his family have lived several years in Finland. His wife Suzi, his late wife Ann, and he have eight children—Kancie, Matt, Emilee, Ryan, Kourtnie, Amber, Jordon, and Brooke.

PART

I

INTRODUCTION

CHAPTER

1

The Strategic Roles of International Assignments in Globalization

Why do some firms compete successfully in the world marketplace and others lose or fail to gain a global advantage? Some analysts argue that strategy is the key: The winners are the ones with the right game plans. Others contend that the key to success is to develop the right corporate structure. Still others claim that meeting the challenges of global business requires technological innovation.

We think that the key to success is developing global leaders. Whether those people recognize or miss global threats or opportunities is a function of their experience and perspective. Increasingly, research is supporting this observation. For example, recent research (Carpenter, Sanders, & Gregersen, 2000) indicates that U.S. firms headed by chief executive officers (CEOs) with international experience perform better than firms whose CEOs lack this experience. Having employees with international experience also enhances an organization's competitive advantage.

Support for the value companies place on global experience is also apparent from several recent executive appointments. Charles Perrin, the CEO of Avon, was chosen in large part because of his global experience at Duracell®. Similarly, General Motors named Richard Waggoner president and CEO after he turned around the company's floundering operations in South America. As the appointments of Waggoner, Perrin, and other executives demonstrate, organizations are increasingly recognizing the value of international experience in shaping global leaders. Likewise, growing numbers of executives identify their international experience as the single most influential force in their development as managers (Carpenter et al., 2000).

International assignments are also the single most expensive per-person investment a company makes in globalizing its workforce, and unfortunately, most firms are getting anemic returns on this investment. To improve their return, managers must understand the best practices, thinking, and scientific research on international assignments. In combination, this material can offer a sophisticated yet practical guide for maximizing the return on international assignments and developing effective global leaders.

STRATEGIC VALUE OF GLOBAL ASSIGNMENTS

In the past, U.S. firms tended not to use global assignments for strategic purposes. In general, people were sent overseas to carry out specific tasks because management felt that the local talent was not up to the challenge. Because of this tactical approach, the strategic implications of the assignment, for the company and the individual, were often neglected. In contrast, leading companies today have developed a more strategic tactic. For them, global assignments serve several important roles: in succession planning and leadership development; in coordination and control; and in technology, innovation, and information exchange and dissemination.

Leadership Development and Global Assignments

One of the key concerns and responsibilities of current CEOs is developing future CEOs and executives. Savvy CEOs know that their company's future strategy will be no better than the quality of the people in the organization. For example, a Beethoven concerto may be beautiful and powerful, but you would not want it played by a beginning piano student. Likewise, in today's business environment, it is madness to think of trying to formulate or implement a strategy without considering the global business environment in which it is expected to function. The "From the Front Lines" for this chapter offers one executive's perspective on the need to develop global leaders.

Currently, most organizations have a serious shortage of global leaders. Consequently, all companies need to make developing globally capable employees a top priority. Fortunately, most human resource executives do recognize this. For example, in a survey of 108 Fortune® 500 companies, senior executives cited having effective global managers as the top priority in achieving international success. When questioned, Jack Reichert, a former CEO of Brunswick Corporation, captured the importance of developing future global leaders:

> Financial resources are *not* the problem. We have the money, products, and position to be a dominant global player. What we lack are the *human* resources. We just don't have enough people with the needed global leadership capabilities.

FROM THE FRONT LINES:

"The Strategic Role of International Assignments"
by Raj Tatta, Partner, PriceWaterhouseCoopers LLP,
and past Chairman of the International Personnel Association;
and John Neylon, Director, PricewaterhouseCoopers LLP.

For PricewaterhouseCoopers (PWC), international assignments are critical to our offering high-quality service to multinational clients across the globe. These international assignments provide important leadership-development opportunities within PWC and are often seen as strategic ways to develop global leaders. The assignments are also important ways to transfer industry/technical knowledge among our global units.

Of our many clients, few capture the unique challenges of these assignments better than our clients in the international oil and gas industry. The leaders who manage international projects in this complex arena must have extensive international experience to manage effectively.

It is no coincidence that Rich Paterson serves as global leader for our international oil and gas group. Rich was carefully groomed for the task, having initially coordinated staffing for a project for the group's largest client in Moscow. Later, Rich moved to Moscow himself to assume the role of project leader. His 4-plus-year assignment in Russia sensitized Rich to the risks of a major global operation and to the intricate cultural issues international assignees encounter. Rich's experience also made it easier for him to recruit managers and partners who would continue to grow the business, attesting to the strategic role these assignments can play within our organization.

Growing the practice in a global market is a tireless effort starting with planting interest among newly hired staff to working with young partners and managers to ensure that they have the right experience to thrive in an environment that is radically different from the one at home. By identifying and mentoring candidates early in their careers, a bond is developed that enhances the fluid movement of staff throughout their careers.

As the global leader for the oil and gas group, Rich still visits Moscow regularly and still encourages U.S.-based managers to identify future leaders for the group. He emphasizes that working within the global community is very important to understanding how to function effectively in a multinational, multicultural business environment.

At PWC, we recognize that close working relationships developed through international assignments are still grounded in personal contact. These personal relationships enable PWC to deal effectively across cultural and language differences and are critical to meeting the fast-changing needs of our multinational clients.

To reinforce the importance of developing a new generation of global leaders, we only need to consider a few recent and significant developments in the global marketplace. Even a superficial review shows them that there are both breathtaking opportunities and threats to leaders in the 21st century. Consider the following:

- Post-9/11 security concerns.
- Formation of the European Union.
- Uncertain economic future.
- Return of Hong Kong to China.
- Establishment of the Economic Union.
- Emerging economies (China, India, Latin America).
- Political instability in Asia and South America.

To effectively formulate or implement strategic plans for the 21st century, managers and executives must be able to focus on the unique needs of local foreign customers, suppliers, labor pools, government policies, and technology and at the same time on general trends in the world marketplace. For an individual, this requires tremendous environmental-scanning abilities just to pick up the information. It requires vast knowledge and processing abilities to categorize and interpret raw data effectively. It requires being able to understand and work well with people from different cultural, religious, and ethnic backgrounds as well as the ability to manage teams composed of cross-cultural members. Managers who fail to develop these skills and organizations that fail to develop global managers risk being irrelevant in the 21st century. Greg Duncan, a vice president at pharmaceutical giant Pfizer, commented from his own experience. "Working abroad is both critical and essential. Like an MBA, it's now part of the pedigree we look for in identifying talent for our organization. In the future, international experience will be the rule not the exception." According to Duncan, there is no substitute for the learning that occurs on an international assignment. He clearly recognizes the value of international assignments in developing global leaders.

Another example is Philip Morris Companies Inc. Philip Morris claims that a key to its growth is that the company knows its global "benchstrength" and views its people as strategic "weapons." Philip Morris has close to 180,000 employees worldwide in 200 countries. Spending time on an international assignment is a common leadership-development technique at Philip Morris.

General Electric (GE) also understands the critical nature of international experience. GE estimates that 25% of its managers need global assignments to gain the knowledge and experience necessary to understand the global markets, customers, suppliers, and competitors the company faces in the 21st century.

Gillette is another innovator in developing global skills among its employees. Beginning in the mid-1980s, the consumer products company started a program to send its best and brightest on assignments around the world. Among the company's objectives in developing this strategy was to enable Gillette to evolve from being simply a multinational company to becoming a truly global organization with managers at all levels and in all locations who could work in a variety of local markets and who were thoroughly familiar with Gillette's corporate and strategic expectations. Today, a whopping 80% of the company's employees live and work outside the United States.

In response to the difficulty finding high-quality people for international assignments, Colgate-Palmolive, in the mid-1980s, began a program of systematically providing opportunities for high-potential managers to work in a variety of global markets. The goal was to develop the leadership skills required for top international and corporate positions. Colgate-Palmolive also has a program that enables entry-level marketing employees to receive international experience and to be on the fast track to a higher level management position from the beginning of their careers with the company.

These are just a few examples of the increasing number of companies where people are seen as key to formulating and implementing corporate strategy and where global assignments play a strategic role in the successful development and globalization of people. Other pioneers in promoting the value of international assignments include Samsung and Motorola, both of which require their top managers to have international experience.

Coordination, Control, and Global Assignments

Global assignments can also play a strategic role in the coordination and control of international corporations. In today's global business environment, the effective coordination and control of units throughout the world is complicated by three factors.

Transportation and Communication Technology. Although people say that the world is getting smaller, it is actually getting bigger. Because of advances in transportation and communication technology, companies are expanding into countries where they have never been and are globalizing faster than ever. We can now phone or e-mail virtually anyplace in the world in minutes or seconds, and we can go anywhere in the world in a matter of hours. Because communication and transportation were more difficult and time consuming in the past, not every place in the world was a part of normal business operations; the whole world has always been out there, but it has not always been relevant or accessible. Today it is. Consider Fluor Corporation, headquartered in Aliso Viejo, California. A $13.5 billion engineering and construction firm, Fluor has more than 50,000 employees in more than 25 countries across six continents. The coordination and control required to run such a network are staggering.

② *Cultural Diversity.* The second factor complicating effective coordination and control of global units is the breadth and depth of cultural diversity among the countries where the corporation operates. This has implications for customers, suppliers, workers, and government relations. For example, what might be construed as a bribe in one country or culture may well be considered business practice in another (Dennis & Stroh, 1993). Management styles can also differ from one country to another, as Will Fraser, former chairman and managing director of Kodak Australasia, discovered when he was transferred to Western Europe:

> I found the Western Europeans, generally, to be quite conservative. They are very, very intelligent, and have awesome language skills, but it was sometimes important to stress the need for more urgent action. I think in Australia we are not quite as analytical, but we do have initiative and we are prepared to get on with things. Europeans in my experience would look at every issue, from every different direction, and often that is important; however, sometimes I found it frustrating that lots of reasons were posed about why we shouldn't go ahead.
>
> Australians are more willing to embrace new ideas and new technology, if they are properly explained. That's not necessarily so in all of the European countries, where the consumer base tends to be somewhat more conservative as well.

③ *Geographic Dispersion.* The third factor in coordination and control of geographically dispersed operations is the resulting potential for conflicting demands from the governments of the countries involved. For example, one country may demand the transfer of a particular technology that the home-country government restricts; technical specialists produced by the subsidiary in one country may be needed in a subsidiary in another country whose immigration policies prohibit their transfer because of their nationality. Such was the case for Nike in China.

When Nike set up a production agreement with a Chinese manufacturing operation to produce running shoes primarily for export sales, the company discovered it needed certain technical expertise in the Chinese contractor's operations. Although China offered substantial savings in labor expenses, local Chinese management and workers lacked the technological know-how to produce the quantity and quality of shoes Nike desired. Nike's most efficient and advanced operations were in Korea, but when Nike tried to send Korean technical specialists in to help the Chinese operation improve its production methods, the Korean workers were denied even temporary visas. This action was primarily because of the supportive political relationship between North Korea and China and the political tensions between North Korea and South Korea. Eventually, to transfer technological know-how to the Chinese operation, Nike officials made tapes of the training and instruction that the Korean workers could not provide in person.

Geographical distance, cultural diversity, and conflicting government demands push firms toward fragmented strategy and operations even while they increase the importance and difficulty of effective coordination and control. Policies and manuals can sometimes facilitate coordination and control, but they are subject to translation, interpretation, and differences in execution often caused by local conditions (culture, government policies, economy, etc.). Furthermore, as subsidiaries grow and mature, resources such as capital, technology, and expertise may not provide sufficient leverage for home-office control. A senior line executive from the Ford Motor Company noted that this coordination and control may be achieved best by using talent from around the world:

> Prior to Ford 2000, we were a collection of regional companies and 99% of the people could advance a very long way in a single region. Two years ago, Ford 2000 flipped the light switch on for global leaders. The need to grow in markets where we didn't have operations suddenly demanded a portfolio of skills in key people that were not what we had developed in a regional setting. Now I have the responsibility to find people for positions around the globe. Guess what? None of these people come from North America or Europe. They're from South Africa, Australia, Taiwan, and New Zealand where they gained the unique skills required to start up overseas operations.

Like maturing children, employees in subsidiaries may want more freedom, and they may resist direction and control from the parent company. Global assignments are an effective means of placing individuals with shared objectives and interpretations in key positions around the world to serve as critical sources as well as means of coordination and control.

Technology, Innovation, Information Transfers, and Global Assignments

Geographical distance, cultural diversity, complex local and global demand or supply conditions, dispersed innovations, and so on create a tremendous need for the various units of an international firm to share and exchange information. Figure 1.1 shows that information can flow to or from the parent organization, in or out of a foreign subsidiary, across foreign subsidiaries, or in none of these directions.

Although each cell in Fig. 1.1 is fairly self-explanatory, all may benefit from a brief description. Although the analysis and descriptions could be applied to any organizational unit within a company (e.g., headquarters, foreign subsidiary, division, or even department), we use a foreign subsidiary as the example.

An *island* foreign subsidiary is one in which little information (e.g., competitive intelligence, product technology, strategic direction) flows **in** from the parent company or other foreign subsidiaries or **out** to these same units. In essence, the subsidiary "does its own thing" in its domestic market.

	Low Flow In	High Flow In
Low Flow Out	Island	Implementor
High Flow Out	Innovator	Integrator

FIG. 1.1. Information flow.

An *implementer* foreign subsidiary is one in which a lot of information flows **in** from the parent company or other foreign subsidiaries, but little flows **out** to these units. In essence, the subsidiary does what it is told to do (i.e., implements) in its domestic market.

An *innovator* foreign subsidiary is one in which little information flows **in** from the parent company or other foreign subsidiaries, but a significant amount of information flows **out** to these units. The subsidiary is a source of ideas, direction, and so forth for other units.

An *integrator* foreign subsidiary is one in which significant amounts of information flow **in** from the parent company or other foreign subsidiaries, and significant amounts of information flow **out** from these units. The heavy two-direction flow of information is what creates the need for significant levels of integration and coordination.

None of these four descriptors is inherently good or bad. Each could be appropriate in a given business environment. The critical point is that except in a totally insulated and isolated subsidiary, the flow of information is an important strategic function. It provides the basis for making strategic and competitive decisions. Mechanisms such as newsletters and intracompany conferences are ways of facilitating information sharing and exchange. Increasingly, the Internet and e-mail are being used for this purpose. Nevertheless, the most valuable information tends to be rich in texture, nuance, and subtleties. As a consequence, it is not easy to digitize. For there to be an effective exchange of this kind of information, there must be a relationship and trust between the parties. Global assignments provide the opportunity for people to work together side by side over an extended period of time, thereby developing the level of trust and understanding necessary for exchanging rich information. Keep in mind, however, that this exchange of information takes place both during and after the global assignment. Although expensive, transferring personnel is necessary. The benefits from relationships developed among the various individuals do not stop after each returns home or goes on to the next international assignment; they continue. Individuals continue to exchange competitive, market, and technological information because of the personal relationships developed during the international assignments. As a consequence, the joint venture is able to respond much more quickly to competitors because more people have more information sooner.

While global assignments can play tactical roles, leading companies also use them to achieve strategic roles. International assignments are vital to the development of global leaders. They are extremely valuable in the coordination and control of a firm's worldwide operations, and they are effective in the exchange and enhancement of technology, innovation, and information transfer—both during and after the time abroad.

RETHINKING OUR VIEWS OF GLOBAL ASSIGNMENTS

Despite the strategic role that international transfers can play in a corporation's ability to succeed in the marketplace, many executives have a rather narrow and myopic view of the value of global assignments and who should be involved in them. The vast majority of U.S. firms select candidates for global assignments primarily on the basis of the technical requirements of the position. Succession planning and managerial development are often irrelevant, not to mention ability to perform effectively in another culture.

Even when executives do begin to focus on the strategic role that global assignments can play, that focus is typically directed toward the assignment of parent-country nationals to foreign subsidiaries. Two factors combine to create this focus on the strategic role that global assignments can play for host-country managers.

First, many foreign governments are pressuring multinational firms that operate within their borders to "localize" management—that is, to develop and promote local managers to positions of global visibility and remove expatriates. Second, many firms have discovered that local managers with no international experience have a tendency to be overly sensitive to local conditions, to view nonlocals as alien, and to misunderstand or even fight corporate directives, plans, and strategy. Host-country managers often become liabilities rather than assets when they do not understand the parent firm, its global strategy, or how other foreign subsidiaries are related to one another. As a consequence, it may not be a good move simply to transfer a manager into an expatriate's position to lower costs or to minimize pressures from local authorities. For this reason, firms have begun to think about and utilize international assignments to globalize people throughout their worldwide operations. For example, Kodak has a program whereby high-potential, host-country managers are identified and then sent on global assignments to important operations in the United States. Such assignments are designed to serve all three strategic purposes described earlier. The overseas assignments facilitate strategic succession planning and managerial development because they provide these managers with experience outside their home countries, broaden their perspective, enhance their knowledge base, heighten personal and communication skills, and improve their ability to assume higher positions back home. The assignments enhance the coordination and control functions of the corporation by socializing the individuals into the Kodak culture and philosophy. They also facilitate information sharing between the foreign managers and domestic U.S. managers.

It is not necessary, however, for a host-country national to be sent on a global assignment only to the parent-firm's country. Ford Motor Company increasingly transfers foreign nationals to "third countries"—that is, countries that are foreign to the individual and to the parent firm. Officials at Ford believe that such assignments enable managers to develop necessary leadership skills for the future, facilitate coordination and control functions between the parent and subsidiaries and among subsidiaries, and enhance the sharing and exchange of information throughout Ford's worldwide network.

In short, although global assignments are most often utilized to "fight fires," they can also be utilized to meet strategic objectives. Although utilizing global assignments for strategic roles is clearly important for parent-country employees, it is equally powerful for globalizing key people throughout the firm's worldwide operations.

COSTS OF POORLY MANAGED GLOBAL ASSIGNMENTS

At this point, some readers may be convinced of the strategic role of global assignments, whereas others may not be. Whether global assignments are used to fight fires or to develop future leaders, it is important to describe the wide-ranging and severe costs that the improper design and execution of global assignments can create. These costs can be roughly divided into five categories: failed assignments, "brownouts," turnover after repatriation, downward-spiraling vicious cycles, and gutting of executive capability at headquarters.

Failed Assignments

The proportion of U.S. expatriates who fail in their global assignments (that is, return prematurely) is significant. Estimates are that between 10% and 20% return early. The failure rates for European and Japanese firms is less than half this rate. (The reasons the failure rates are so high among U.S. corporations as well as differences between the failure rates of U.S. and European or Japanese firms are examined in more detail in chap. 5, this volume.)

Direct Moving Costs. The first and most direct costs of failed assignments are those associated with physical displacement. It is very expensive to ship an international assignee and his or her family and belongings overseas (65% of U.S.

assignees are married; Windham International, National Foreign Trade Council, SHRM, 2002). For example, a typical transfer to Tokyo from the United States might cost $100,000 (including a relocation allowance, temporary living costs, brokers' commission, one-way travel to Japan, moving costs, and money for property management). It would cost approximately another $60,000 to $70,000 to bring the international assignee and family home plus another $100,000 to send a replacement. The relocation costs alone can be more than $250,000 for a failed assignment.

Downtime Costs. During the expatriate's predeparture preparations, there is a period when the individual is receiving a full salary but is not capable of fully performing duties. There is also a time during the first few months of the global assignment when the expatriate is adjusting to the new culture, the environment, and the job. This is natural. In the case of failed assignments, the problem is twofold. Unlike expatriates who recover, adjust to, and perform well in overseas positions, those who fail provide no long-term return on the downtime. Also, once an expatriate is overseas, the base salary, foreign-service premium, housing allowance, education allowance, cost-of-living differential, tax-adjustment allowance, and so on usually at least double the total compensation package; and this doubles the cost of the adjustment downtime for which there is no long-term return in the case of a failed assignment.

Indirect Costs. In addition to these measurable economic costs, there are significant indirect, difficult-to-measure costs to both the organization and the individual. In the case of the organization, a failed global assignment can result in damage to several important constituencies including local national employees, host government and local suppliers, customers, and community members. If these damages occur, the replacement person, even the most capable, may find it hard to repair them and effectively carry out other duties and tasks. Despite the difficulty of quantifying these costs, they are sometimes the most significant. For the individual, a failed global assignment will likely hurt his or her career and self-esteem. International assignees we have interviewed over the years have said that they stayed in their global assignments only because they feared the negative consequences of leaving early. For the firm, lost business opportunities, lower employee morale, upset government officials, and disappointed customers can all negatively affect financial performance.

Brownouts

Brownouts are managers who do not return prematurely but are nevertheless ineffective in performing and executing their responsibilities. Estimates are that 33% fall into this category (Black & Gregersen, 1999).

Research shows that the costs associated with expatriate failure range from $300,000 to $1 million per assignment (Black & Gregersen, 1999). These costs reflect only directly identifiable expenses such as compensation, training, orientation, development, and termination. The figures could easily be twice as high if they included hidden costs associated with failure such as damaged relationships, decreases in productivity, resignations by host-country nationals, missed business opportunities, and so on. Perhaps the best way to illustrate some of the costs associated with poor performance during the international assignment is through an experience at GE.

Case 1: General Electric—Cie Générale de Radiologie

In the mid-1980s, GE went through a massive strategic restructuring during which medical technology became one of GE's core business areas. GE's stated goal was for all strategic business units (SBUs) to be in first or second place among their worldwide competitors.

In an effort to increase its global strategic position in medical technology (especially imaging technology), GE took control of Cie Générale de Radiologie (CGR) in 1988. CGR was a French medical equipment manufacturer that was owned by the state and run much like a government ministry. GE got CGR and $800 million in cash from state-controlled Thomson S.A. in return for GE's RCA consumer-electronics business. The acquisition of CGR was viewed by many as a brilliant strategic move, and GE projected a $25 million profit for the first full year of operations. Things did not go as the strategic planners had projected, however.

One of the first things GE did was to organize a training seminar for the French managers. GE left T-shirts with the slogan "Go for One" for each of the participants. Although the French managers wore them, many were not happy about it. One manager stated, "It was like Hitler was back, forcing us to wear uniforms. It was humiliating."

Soon after the takeover, GE also sent U.S. specialists to France to fix CGR's accounting system. Unfortunately, these specialists knew very little about French accounting or reporting requirements, and they tried to impose a GE system that was inappropriate for French reporting requirements and for the way CGR had traditionally kept records. This problem (and the working out of an agreement) took several months and resulted in substantial direct and indirect costs.

GE then tried to coordinate and integrate CGR into a Milwaukee-based medical equipment unit. Because CGR had racked up a $25 million loss instead of the projected $25 million profit, an executive was sent to fix things at CGR. Several cost-cutting measures, including massive layoffs and the closing of roughly half of the 12 CGR plants, shocked the French workforce. The profit-hungry culture of GE continued to clash with the state-run uncompetitive culture of CGR.

> *GE's efforts to integrate CGR into the GE culture by putting up English-language posters everywhere and flying GE flags met with considerable resistance from the French employees. One union leader commented, "They come in here bragging, 'We are GE; we are the best, and we've got the methods.'" The reaction was so strong that a significant number of French managers and engineers left GE–CGR, and the total workforce shrank from 6,500 to 5,000. GE officials had estimated GE–CGR would produce a big profit in 1990; instead, it lost another $25 million.*

We do not mean to point an accusing finger at GE; most experts view GE as a well-run multinational corporation. However, the example does illustrate the costs of poorly designed or poorly executed global assignments. Managers who do not perform well abroad often do not provide an adequate return on investment; they may initiate programs or projects that cost time and money, and they may damage relationships that are difficult to repair. They may even drive out high-potential local managers who will be needed in the future.

Turnover After Repatriation

Repatriation is perhaps the least carefully considered aspect of assignments and potentially the most costly. According to the *Global Relocation Trends Report* (Windham International, National Foreign Trade Council, & Society for Human Resource Management, 2002), approximately 25% of managers leave their companies within 2 years after repatriation. On average, firms spend $150,000 to $250,000 on all expenses (salary, allowances, etc.) for each overseas manager each year (Stroh, Gregersen, & Black, 1998). For example, maintaining an American expatriate who earns a base salary of $100,000 and has two children costs at least $220,370 in Tokyo, $180,312 in Singapore, and $157,762 in Beijing per year. Philip Morris claims it spends $816,000 per year to maintain a $100,000 salaried expatriate in Tokyo ($25,000 targeted bonus, $86,000 commodities/services allowance, $5,000 hardship pay, $171,000 housing costs, $31,000 home leave, $90,000 education for two children, $75,000 relocation costs, plus $125,000 salary equals $816,000). For the average 4-year assignment, the firm invests from $1 million to nearly $3.5 million per manager. This investment is especially important if the assignment is part of succession planning and the managerial development agenda. On average, U.S. firms face a 20% chance that they will receive no long-term return on this substantial investment. They may even have to duplicate these investments to develop replacements. Furthermore, because these employees often leave to join competitors, the firm has de facto financed the development of global leaders for its rivals. The factors that contribute to repatriation adjustment and turnover problems are explored in detail in chapter 9 (this volume); the important point here is the tremendous expense that the poor repatriation management can produce.

Downward-Spiraling Vicious Cycles

All these costs can combine to create circumstances that set off a downward-spiraling cycle, which may erode or destroy a firm's global competitive advantage. Failed global assignments, rumors of brownouts, and turnover problems among repatriates can lead the best and brightest in an organization's worldwide operations to view global assignments as the kiss of death for their careers. This reputation makes it difficult to recruit and send top-quality candidates on global assignments, which in turn increases the likelihood of more failures. This downward-spiraling quality of candidates and performance can feed on itself, gaining momentum with every turn. The firm may stop sending anyone on assignments outside the home country, which in turn can lead to more difficult coordination and control as well as to problems exchanging information. Like a plane falling from the sky in a tailspin, it's difficult to regain control once this spiral starts. Recovery is possible, however. GE Medical, for example, learned from its experience in France, dramatically changed its policies and practices, and is now one of the leading companies in terms of international assignment policies and practices.

Gutting of Executive Capability at Headquarters

Perhaps most important, this vicious cycle can lead to a shortage of leaders who have vital understanding and experience in the global arena, which can result in poor strategic planning and implementation and an ever-worsening competitive position globally. Mismanagement ultimately leads to a loss of managerial resources—a fatal flaw for globalization—because without international experience, executives are unable to formulate and implement global strategy accurately. As growing numbers of upper echelon managers gain international experience and recognize its value, executives will undoubtedly expect managers moving up the ranks to have such experience themselves.

Many managers and executives may feel that such a scenario is unlikely; but this cycle is easier to start than might be expected, and it is much more difficult to stop or reverse than might be estimated. To fulfill the strategic roles that global assignments can play or simply to avoid the staggering costs that poorly managed assignments can produce, organizations must provide a framework for effective people management and international assignments.

Framework for People Management and Global Assignments

Before we explore how to manage global assignments and the people in them successfully, it is essential to develop a basic framework for our discussions and recommendations. From our perspective, the effective movement and management of people in global assignments are most logically framed in terms of the generic issue of people management. The term *people management* is often used as if ev-

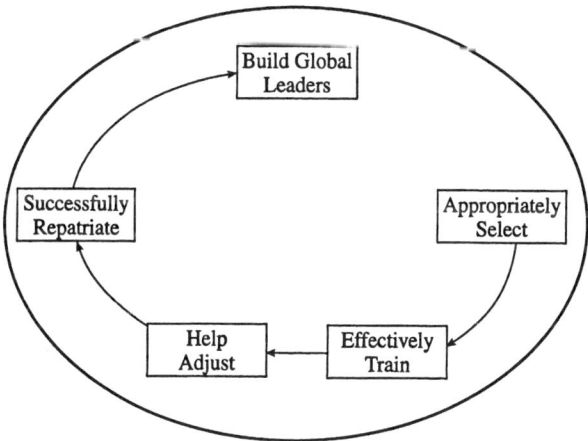

FIG. 1.2. Global assignment success cycle.

eryone had the same idea of what it means, but we all have very different ideas of what is involved. Figure 1.2 illustrates conceptualization of the term through the global assignment success cycle.

FIVE DIMENSIONS OF PEOPLE MANAGEMENT

We view people management not as a function of a specific department (such as personnel or human resources) but as a set of activities that any manager in any functional area of a firm must master. Each activity builds on the others as the process becomes an integrated package. Simplified, there are five generic functions of managing people (Copeland & Griggs, 1985).

Getting the Right People (Recruiting/Selecting). First, a manager must identify, recruit, and then place individuals in appropriate positions within the organization. Sometimes this process involves people who are already in the company, and sometimes it involves hiring individuals from outside. This first aspect of people management also includes determining the types and numbers of individuals who will be needed in the future for certain positions and examining the existing pool for people who could meet those needs. At the corporate level, this raises such questions as "Are managers with the necessary skills and experience placed in strategic positions throughout the firm's global operations?"

Most U.S. multinationals select individuals for overseas assignments on the basis of their domestic track records. Although this is one important criterion in selecting employees for global assignments, it does not guarantee success overseas. Research has demonstrated that several factors need to be examined including personality traits and the ability to adapt to cultures different from the one at home. In chapter 3 (this volume), we detail the problems that firms face in selecting candidates for global assignments and discusses common practices, their consequences, and the principles that explain why some practices are more effective than others; we also describe how firms can successfully structure their international staffing and selection practices.

Helping People Do the Right Thing (Training). The jobs that managers are expected to do and the standards by which performance will be judged must be determined, and the necessary training must be provided. These requirements beg the question of whether a firm's managers—who must deal with employees, customers, suppliers, or competitors from different cultures and countries—are adequately trained to understand and work successfully with these groups. The vast majority of U.S. multinational firms fail to prepare individuals adequately to work with individuals from other cultures. Although a 2002 survey said 64% of companies provide some type of cross-cultural preparation, approximately 70% of Americans who must work overseas receive inadequate training and or preparation for their international work (Windham et al., 2002). In chapter 4 (this volume), we detail common practices and consequences in corporations and the dynamics that explain why training is or is not effective; we also describe how effective training programs for global assignments can be designed and implemented.

Determining How People Are Doing (Appraising). Once an employee has been trained, his or her performance must be measured. However, measuring the job performance of managers who have been sent overseas is tricky. For example, if traditional U.S. variables such as profit, sales, and market share are utilized as quantitative measures, should factors that apply to the local business environment (such as movements in exchange rates) be incorporated into the performance evaluation as well? Most firms have little idea about which factors facilitate or inhibit cross-cultural adjustment, organizational commitment, or job performance during an overseas assignment. In chapters 5, 6, and 7 (this volume), we detail the important factors and the underlying processes that affect these outcomes; in these chapters, we also describe effective means of enhancing and measuring adjustment, commitment, and job performance.

Encouraging Specific Behavior (Rewarding). In addition to measuring an individual's job performance, the organization must provide rewards for specific performance behavior as well as general compensation and benefits. This aspect of people management raises such questions as "Should all managers receive equal benefits and bonuses regardless of the country in which they are working?" Most firms complain about the high costs of global assignments, and many firms have cut the total number of international assignees to reduce costs. In the absence of understanding or analysis of the reward systems for global managers, this is unlikely to improve the performance of individual managers. In chapter 8 (this volume), we detail common practices and consequences in the area of compensation and rewards and discuss the dynamics that motivate global managers; we also describe how rewards can be structured to enhance motivation and performance during global assignments.

Doing Right for People (Developing). Over the long term, a sequence of positions, opportunities, responsibilities, and so forth will be needed to develop the maximum potential of managers. This aspect of management raises such questions as "How should managers sent on international assignments be utilized once they return?" There is strong evidence that most U.S. firms do very little planning for the systematic development of global managers. Specifically, most U.S. firms do little planning for the return and integration of global managers who have been overseas for some time. Many of these employees—as many as 75%, according to one study (Black & Gregersen, 1999)—feel they have been demoted in their new positions after returning from an assignment abroad. In chapters 9 and 10 (this volume), we detail the factors and processes that affect repatriation adjustment, commitment, and turnover; we also describe how firms can more effectively manage this aspect of global assignments. In chapter 11 (this volume), we combine the separate aspects of successfully managing people and global assignments into a comprehensive system and place it in the context of the likely business environment in 2005 and beyond.

Figure 1.3 provides a list of common questions about the five aspects of people management and international assignments that most companies have to face and answer.

PEOPLE MANAGEMENT IN THE GLOBAL CONTEXT

Although most of the principles and best practices are relevant to all firms independent of their size or industry, some important nuances are beneficial to explore. These nuances are perhaps most apparent when we look at a firm's current and future globalization plans.

Staffing	• What characteristics should be utilized in selecting expatriates? • Why are expatriates likely to accept/reject international assignments? • How can a larger pool of potential international assignees be identified and developed? • Can questionnaires and interviews be used effectively to screen and select candidates who are more likely to be successful overseas?

FIG. 1.3. *(Continued).*

Training	• Should all expatriates receive training? • What are the most effective training methods for international assignments? • What should be the relative mix and content of predeparture and postarrival training? • How can the cost effectiveness of training be calculated?
Appraising	• How can expatriate performance be monitored in light of exogenous factors, such as foreign exchange rates, that can dramatically affect business performance? • What is the appropriate mix of quantitative and qualitative measures of expatriate job performance? • To what extent should local nationals be involved in evaluating expatriates?
Rewarding	• To what extent should an expatriate compensation package be utilized to entice employees to accept overseas assignments versus simply equalizing cost-of-living expenses? • How should different tax laws for expatriates from different home countries be factored into compensation? • How should compensation inequities be handled that arise when expatriates of similar level but from different home countries work in the same assignment country?
Developing	• To what extent should shorter assignments be used to develop young managers early in their careers? • How can the immediate task needs in the assignment and the development needs in the individual be effectively balanced? • How can employees with successful international experience be effectively repatriated into other organizational units?

FIG. 1.3. Basic questions of people management and international assignments.

Patterns of Globalization

Although the five aspects of people management apply to all firms engaged in transferring people across national borders, slightly different issues emerge as a function of the stage or pattern of a firm's globalization. When we use the term *globalization,* we have in mind not one fixed point at which a firm is globalized but a series of patterns of globalization. In the next five sections, we briefly describe each general pattern of globalization and some of the specific international-assignment challenges and issues associated with each.

Export Firms. Export-oriented firms have most of their value-chain activities in one or a few countries and have little coordination among global organizational units. Firms of this type, such as L.L. Bean, are concerned primarily with exporting products and then marketing, selling, and distributing them in several countries. From a strategic perspective, export sales are an "add on" to domestic sales. The best evidence of this can be seen when demand exceeds supply. In this situation, the demands of domestic customers are given priority over those of international customers. Consequently, the focus of international people-management activities is often staffing, training, appraising, rewarding, and developing local nationals involved in these downstream activities. There is a generally low use of expatriate managers in export-oriented firms; instead, home-based managers with geographical or product responsibilities visit various international countries and sites. When international managers are used at all, they are usually placed in positions of general management and have rather broad geographical responsibilities. Most expatriates are sent out from the country of the parent firm, and very few if any managers are transferred from foreign countries into the parent organization or between foreign operations.

In firms in which international operations are export oriented, international assignments and assignees are often not top priorities. Careful attention to best practices, policies, and top international candidates is lacking. In one sense, this is both understandable and logical. However, problems arise when the firm begins to make international activities a stronger priority. At this point, the firm may send its best and brightest out to its now more strategically important international operations. Unfortunately, the minimum attention on international assignments and assignees makes this action a difficult transition. Because the firm has not valued international assignments, the best and brightest may be reluctant to go "out of sight and out of mind." Executives in export-oriented firms today must balance the needs of the moment with laying the groundwork for the future.

Multidomestic Corporations (MDCs). A multidomestic corporation is one that has multiple foreign operations, but each foreign affiliate is basically focused on local markets, competitors, and so on. This situation results in multiple domestically focused operations. Such multidomestic firms as Procter & Gamble, Honeywell, and Alcoa are most common in industries in which competition in one country (or in one small group of countries) is independent of competition in others. Because competition for each geographically distinct unit within the firm is focused on the country and the market in the unit's location, a high degree of specialization and adaptation of the unit's value-chain activities is required. The culture- and country-specific knowledge that local nationals have is important for the appropriate specialization of the unit's activities. It is therefore only natural that the use of international managers, although higher than in the case of export firms, is relatively low in MDCs. Moreover, the international managers tend to be of two types: executives or technical specialists. The executives are often from a cadre of "career internationalists"—managers who made and spent most of their careers outside the firm's home country. The technical specialists are generally on assignment for a relatively short time (1 to 2 years) and are overseas for specific reasons (e.g., to transfer a particular technology or to solve a particular problem).

The multidomestic orientation generally does not lead to systematic policies and practices formed for expatriate selection, training, or repatriation because only a small percentage of employees ever serve in overseas assignments, and these employees are generally out of sight, out of mind insofar as policymakers at the home office are concerned (Windham et al., 2002). Instead, MDCs focus on international-assignment compensation practices and policies to attract a sufficient number of reasonably capable international managers. However, from a strategic perspective, MDC firms should focus on all aspects of international assignments. International-strategy formulation and implementation are primarily restricted to an individual country, and for the country centered strategic plan to be effective, key executives must incorporate the important and unique aspects of the country, culture, and market where the unit competes into their business plans. Consequently, even though only a few international managers are utilized in MDCs, great care must be taken to ensure that the right people are selected, trained, and prepared for a quick and effective adjustment so that they can make exceptionally good decisions about what and how things should be done in the local market.

Global Strategy Multinational Corporations (MNCs). Like MDCs, global-strategy MNCs (referred to simply as MNCs) also have geographically dispersed operations and units. In contrast to multidomestic firms, however, MNCs such as Monsanto, General Motors, and IBM have extensive coordination among units in different geographical locations. This coordination and control is achieved through a variety of mechanisms. One particularly effective means is the establishment of a common organizational culture across worldwide operations. MNCs tend to utilize more international managers, utilize both home-country and third-country international managers, and place international managers in a variety of organizational levels within foreign operations.

In addition to engaging in the common practice of sending home- or parent-country nationals on overseas assignments in foreign countries, MNCs often have foreign nationals serving in overseas assignments at the home or parent office. Bringing foreign nationals into the home or parent country (sometimes called *inpatriation*) can help socialize foreign nationals to the philosophy or culture of the parent firm as well as cross-fertilize the home operation with the ideas of foreign nationals and their own home-country operations. Because the international movement of employees provides a powerful informal means of controlling and coordinating activities across countries, MNCs with this orientation are more likely to have developed, over time, systematic policies and practices for expatriate selection, training, and repatriation.

The two-way directionality of informal coordination raises some very important issues with respect to repatriation. Because expatriate employees gain important experience that prepares them to understand various international markets, competitors, and operations, they are in a unique position to contribute effectively to international-strategy formulation and implementation at corporate and regional headquarters. Therefore, in addition to a need for excellent selection, training, and international-assignment support policies and practices, MNCs also need effective repatriation policies. Without them, the typical high postrepatriation turnover will significantly and negatively impact the firm's ability to populate its management ranks with global leaders with the requisite international knowledge and experience the company needs to compete effectively.

Multifocal Corporations. Firms at the multifocal stage of internationalization, such as Coca-Cola and Airbus, are generally within industries in which a firm's position in one country is substantially influenced by its competitive position in another. The objective for a firm with a global orientation is to coordinate value-chain activities on a global scale and thereby capture comparative advantages and links among countries. If transnational or global corporations were free from government restrictions, they would tend not to think of employees as home-country managers or local nationals; they would simply try to place individuals with comparative advantages and needed skills in appropriate locales much the way a firm moves people within domestic borders. These value-chain activities could be concentrated in specific countries with comparative advantages relative to other countries. For example, if research and development (R&D) were in France, scientists and related people with the knowledge and skills to contribute to R&D would be moved to France regardless of nationality. The reality is that a "borderless world" in terms of the movement of people does not yet exist. Nations have visa laws, domestic employment policies, tax structures, and other restrictions that prohibit the free and uninhibited movement of people across borders. In practice, global corporations are restricted in the movement of international staff people.

Even though global companies are not completely free to move people across borders, they still must focus on developing managers and executives with global leadership capabilities independent of nationality. As a consequence, these firms must benchmark their international assignments against the best policies and practices not just within their industry or home country but against the best firms across the globe.

The Issue of "Fit." No matter which globalization stage a firm is in, its international assignment policies and practices must fit the environment and must be congruent with each other. The balance of international assignment policies and practices with the marketplace creates an *external fit.* The balance and congruence among the five aspects of people management within a firm (staffing, training, appraising, rewarding, and developing) constitutes *internal fit.* Without good external and internal fit, a firm will experience significant difficulties in effectively formulating and implementing strategies. Two examples will illustrate this point.

Case 2: WestCoast Bankcorp

A large west coast bank (the name of the firm has been disguised) was trying to reduce its total expenditures on international assignments. The bank was at a coordinated MNC stage of globalization. It had operations in more than 20 countries and sought to closely coordinate among its subsidiaries such value-chain activities as lending, foreign exchange, and retail banking. The bank also sought to compete by offering low interest rates and fees for the services it provided.

The bank found that its international managers were usually two to three times more expensive to employ than comparable employees in the home country and several times more expensive to employ than local nationals. These high costs for expatriates were consistent with what most other U.S. banks experienced. WestCoast Bankcorp tried to lower its costs by reducing its aggregate expatriate costs. This goal was achieved over just a few years by reducing the total number of global managers by half. The bank also tried to reduce the average cost per expatriate by reducing or cutting various aspects of the standard expatriate package (reducing allowances, locating global managers in low-tax countries, cutting predeparture training, etc.).

The critical implication of this particular cost-reduction program concerns the issue of external fit. The key question is whether the effort to lower total costs was carried out in a manner that inhibited or facilitated the firm's competitive advantage. Although cutting the total number of global managers reduced the aggregate costs, coordination and control between the parent and subsidiaries as well as among subsidiaries became much more difficult. The effort to lower the cost per global manager by reducing predeparture training also reduced these managers' ability to function effectively and made it difficult to attract top-quality managers to international assignments. Fewer global managers were now being placed in top positions within the foreign subsidiaries, and to facilitate effective coordination, they had to understand both the home office and the local operation. Lack of resources for cross-cultural training reduced their ability to understand the local situation and to facilitate coordination.

This couldn't have happened at a worse time. WestCoast Bank found itself competing with new foreign rivals, especially from Japan, in its home and in its key overseas markets. WestCoast Bank needed global coordination and control to fight off the coordinated attacks from foreign competitors, but its international-assignments practices and policies made it difficult to do so. This poor external fit had a significant and negative effect on the bank's subsequent financial performance.

Case 3: International Hotel

International Hotel is a large U.S.-based hotel corporation that tried to maintain its competitive advantage through differentiation by offering guest services (e.g., business centers equipped with fax machines, computers, copiers, secretaries, etc.) that far exceeded those of its competitors. The exact services varied considerably from country to country as a function of its multidomestic pattern of globalization. International managers were expected to formulate their own location-specific approaches to this "high-quality service" strategy and were monitored on the basis of their financial results. Reasonable levels of predeparture and postarrival training were provided, which facilitated the managers' ability to quickly gain an in-depth understanding of the values and needs of local markets and to design services to meet those needs.

Unfortunately, much of an international manager's bonus in this organization was based on overall corporate results, so whereas the training led managers to want to focus on understanding local markets, the appraisal and compensation policies had the opposite effect. The managers knew that if they could reduce costs (in some cases by eliminating certain costly services), profits would increase and so would their bonuses. In this case, the internal fit among training, rewarding, and appraisal was poor and did not facilitate or encourage expatriate managers' ability to differentiate their hotels in the local markets where they were competing.

Key Implication 1: Figure and Ground

As anyone who has ever seen a photograph or painting knows, what you focus on as you look at the picture depends on what is figure and what is background. If a tree is front and center in the picture, you tend to focus on it and see it in detail. If the tree is part of the background, you might not even notice it. Managers in your company are astute observers of figure and ground. On many occasions, we have been called in to help a company figure out why it cannot get good (let alone the best) managers to take international assignments. Although a variety of factors can influence a decision to accept or reject an overseas stint (we explore these in depth in chap. 3, this volume), time and time again employees have said something similar to the following:

Yeah, I hear the rhetoric about international, but talk is cheap. If international is so important, why don't any of the senior executives have any overseas experience? Why, if it matters so much, was (Fred, Sally, Juan, or whoever) left dangling in the wind? They had only a month's notice before they had to leave. They got zero training. They were out of sight and out of mind while they were gone. No one knew what to do with them when they came back. Their peers moved ahead of them during the assignment. So why won't I or others like me take an international assignment? I wonder.

Quite frankly, when it comes to international assignments, many companies do not know what their employees perceive as figure and what they perceive as ground. Many firms have been surprised by the results of surveys and interviews with their people. However, the insights proved invaluable in knowing where to focus efforts to improve both the perception and the reality of international assignments.

Key Implication 2: The Whole Picture

The best results are achieved by looking at a whole picture. Traditionally, compensation has been the focus for most companies. In one sense, this is understandable; after all, people have to get paid and taxes have to be taken care of. However, leading companies use a more systematic perspective. They know that all aspects of international assignments—selecting, training, rewarding, appraising, repatriating—have to be considered together. Each can have a significant impact on the others. For example, if you fail to select the right person, no amount of training can compensate for that. However, starting with the right person but ignoring training will also not guarantee success. Whether in music, sports, or business, even the most talented individual needs education, training, and coaching to excel. So, even though we examine each of the aspects of international assignments separately, they are all interconnected and best understood by looking at the whole picture.

Key Implication 3: Looking Ahead

Even if we look at the whole picture and get the right things in figure and background, as the picture is today, we must also look ahead. Global leadership skills are not developed overnight, but the need for these skills can become apparent with little notice. If a foreign competitor starts moving in on your home turf or major customers expand internationally (or a new business opportunity appears in an emerging market), your firm will be in desperate need of high-quality global leaders in short order. Yet finding such leaders is one of the greatest challenges companies face today (Varma & Stroh, 2002). The major reason is that in most companies international-assignment policies and practices lag behind the strategic needs of the business. Executives need to look 5 or 10 years out into the future, anticipate the quantity and quality of global leaders they will need, and start building immediately.

As we said at the outset, international assignments are not the only means of developing global leaders, but they are the most powerful. Consequently, they must be the best planned and best managed of a firm's international activities. In too many companies, this is not the case. To gauge where your company is, ask yourself, if I asked 20 senior executives to rank the importance of the following international activities, where would international assignments fall: international strategy, organizational structure for global reach, international marketing, offshore manufacturing, cross-border alliances and joint ventures, and international assignments?

International assignments should rank at the top—not for altruistic reasons but for hard-core business reasons. For example, although executives with international experience are still in the minority, those organizations that employ such executives appear to have a competitive advantage (Carpenter et al., 2000). Unfortunately, developing tomorrow's global leaders who have international experience does not happen overnight; if your firm will need them in the future, it needs to start developing them today. Few things could be of greater value than (a) a top-to-bottom review of your current international assignment policies and practices, especially as perceived by the employees of your company, and (b) working to ensure that your overall system has superior external and internal fit.

SUMMARY

We began this chapter by saying that people are the key to global competitiveness and by identifying three strategic functions of international assignments in the global marketplace. Global assignments can play a strategic role in succession planning and leadership development, in the coordination and control of international operations, and in technological and information exchange between the parent company and subsidiaries and among subsidiaries. We argued that in addition to considering global assignments as more than a tool for fire fighting, firms should send both home-country nationals and foreign-country nationals on global assignments. We also pointed out that independent of the fire fighting or strategic role of global assignments, companies often accrue significant and sometimes devastating costs if the assignments are poorly design and poorly managed.

In this chapter, we also presented a general framework of people management to structure later chapters on the selection, training, cross-cultural adjustment, performance, evaluation, compensation, and repatriation. These issues become increasingly important as firms move from export patterns to coordinated MNC patterns of globalization. Consequently, executives need to examine the external fit among the marketplace, the pattern of globalization, and the firm's international management policies.

In the remainder of this book, we summarize the best thinking, practices, and scientific evidence available on the management of people in the international marketplace. In the remaining chapters, we also discuss ways a company can modify general recommendations concerning international assignments depending on the firm's pattern of globalization.

In our style and approach, we try to walk the razor's edge. On one hand, we present the latest scientific evidence and research findings. Like a practicing doctor, we believe you can serve your "patients" best when you have the latest research available to you. On the other hand, we wanted the information to be accessible and easy to read. We wanted to make this book informative but interesting. Our hope is that you will be pleased with the results.

CHAPTER
2

The Process of Cross-Cultural Adjustment

If adjustment to life and work in other cultures were less difficult, far fewer managers and their families would return early from global assignments, and more global managers would perform effectively. The reality, though, is that most people find it difficult to adjust to life in a culture different from the one they're used to. In this chapter, we focus on some of the reasons making cross-cultural adjustment is so often a challenge. In later chapters (this volume), we discuss how adjustment affects such job-related matters as turnover and performance. Before we tackle the topic of cross-cultural adjustment, take a look at Fig. 2.1. It should help you put U.S. and other cultures in perspective. You might be surprised to find out what a small proportion of the population your demographic group happens to be. For example, only 1 in 100 have a college education.

WHAT CULTURE IS AND ISN'T

People usually think of a country's culture as its ceremonies, clothing, historical landmarks, art, and food—that is, things people can see. However, although observing what people wear, eat, and so forth is a first step in understanding a country's culture, it is just a first step. To adjust to a culture, you also need to understand why people dress the way they do, eat the foods they do, and appreciate certain art and music. You also need to understand what they believe and value. In other words, you need to understand both the explicit and the implicit aspects of the culture.

One way to picture a culture is as a tree, with parts visible, above the surface, and parts below—including supporting roots (see Fig. 2.2). The tangible aspects

Why a New View of Culture Is Necessary

If you shrank the world to the size of a 100-person village but kept the proportion of each cultural group the same as in the world today, this is what your village would look like:

It would have 58 Asians, 12 Africans, 10 Europeans, 8 Latin Americans, 5 North Americans, and 7 people from other areas.

Seventeen people would speak Mandarin Chinese, 9 English, 8 Hindi, 6 Spanish, 6 Russian, and 4 Arabic. The other 50 would speak any of more than 200 other languages.

Thirty-three people would be Christian, 18 Muslim, 6 Buddhist, 3 Jewish, and 5 atheist. The remaining 35 would identify with other belief systems.

Twenty people would make 75 percent of the total income; another 20 would make 2 percent of the total.

One-third of the villagers would have access to clean drinking water; two-thirds would live in substandard housing; and only seven would own a car.

Of the 67 adults in the village, half would be illiterate; only one would have a college education.

FIG. 2.1. Why a new view of culture is necessary. *Source:* R. Bahner (2003). Presentation to the International Personnel Association, October 12.

FIG. 2.2. A view of culture.

of the culture—those things one can see, hear, smell, taste, or touch—are the *artifacts,* or manifestations, of the underlying values and assumptions that the people share. These artifacts represent only a small fraction of the country's culture. What we cannot see—the values, assumptions, and beliefs—support and give life to the culture.

In every country, values as well as assumptions and beliefs are passed from older to younger generations forming the foundation of the culture. These values exert a strong influence on the country's artifacts, but more important, these values directly affect people's behavior. To adapt well to a new culture, international assignees need to understand the culture's underlying values and how these values influence what is considered appropriate behavior (Dowling, Schuler, & Welch, 1997; Hoecklin, 1995; Hofstede, 1980; Linton, 1995; Nemetz & Christensen, 1996; Welch, 1994).

By now you may be wondering, how and why do values emerge? How and why do values become widely shared? How and why do values get passed from one generation to another? The basic answer to all these questions is quite simple. All societies face and must solve a specific set of problems. They must figure out ways to communicate and to educate, feed, clothe, and govern their people. Societies experiment with different ways to address these challenges. Methods and ideas that prove successful are kept; those that fail are discarded. In turn, successful methods and ideas are passed on to future generations. Because these fundamental elements of the culture reside in people's heads, they are intangible. However, these intangible elements—the culture's values and beliefs—are what guide and shape behavior we can see. Thus, understanding the intangible elements of a culture is critical to successful cross-cultural adjustment.

Put in different terms, a culture's values and beliefs are like mental road maps and traffic signals. The maps tell the people in the society what the important and valued goals (key destinations) are and the ways to get there; the traffic signals tell them who has the right of way, when to stop, and so on. Like an experienced driver, people retain these maps and traffic rules in their heads over time. People obey the most basic rules without thinking about where they came from or the consequences of not following them. They know what the consequences are for turning right on a red light when it is against the rules, so they wait for the light to turn green.

However natural these maps and traffic rules might seem to a "local," they can be a nightmare to a visiting foreigner. Imagine being put into the heart of the Tokyo freeway system with no map, no road signs, and no idea of the rules governing traffic. Suddenly, your previous road maps and traffic rules are practically useless or perhaps even a threat to your safety—after all, the Japanese, like the British, drive on the left side (Americans would say the "wrong side") of the road.

A similar process occurs as people relocate from one culture to another. Suddenly, previous social and interpersonal guides are useless or even life threatening. The only way to succeed socially and professionally is to learn new maps and rules to navigate social and business situations. The complication is that unlike actual road maps and traffic rules, no convenient manuals explain the deep-rooted beliefs and values of a culture.

What happens when you violate (probably unintentionally) cultural traffic rules? Just as violating some traffic rules will get you only a warning or a small fine, whereas other transgressions may land you in jail, the penalty for violating a cultural rule depends on the importance of the rule that is broken. A helpful way of thinking about the importance of cultural rules and the severity of the consequences for breaking them is to conceptualize cultural values along two dimensions: the extent to which they are widely shared among group members and the extent to which they are deeply held. This conceptualization is illustrated in Fig. 2.3.

Those assumptions, values, or rules of the culture that are widely shared and deeply held generally result in substantial rewards or punishments if those assumptions, values, or rules are broken. For example, a widely shared and strongly held rule in the United States is that people do not talk to themselves constantly or loudly. When we see a person doing this, we become nervous and concerned, even if the person poses no physical threat to anyone.

What happens to people who violate this rule? In many cases, they get sent to mental institutions.

What about rules that are deeply held but not widely shared? In this case, the rewards or punishments are often informal. For example, some people, but not everyone, consider burping after a meal to be a serious violation of proper behavior. You will not be put in jail for burping, but you may be cut out of particular social circles, at least in the United States. In other countries, such as China, you may offend a host by not burping after dinner.

In the case of rules that are widely shared but not deeply held, violations often entail uniform but rather mild punishments; infrequent violations may carry no

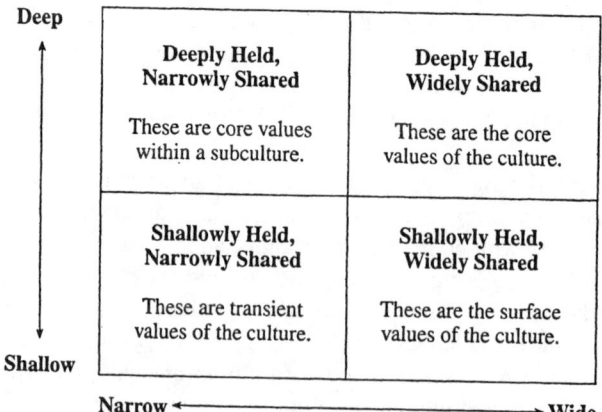

FIG. 2.3. The matrix of culture.

punishment at all. For example, in the United States, people are taught not to interrupt. However, if people occasionally interrupt, it is unlikely that the punishment will be significant.

FUNDAMENTAL CULTURAL ASSUMPTIONS

To understand another culture, we must understand the roots of that culture, which are found in its fundamental assumptions. From these assumptions grow the culture's values, beliefs, and visible artifacts and behaviors. Not all cultural trees have the same root systems. Consequently, it is important to discuss briefly the nature of assumptions and how they differ from one culture to another. Fortunately, we do not have to examine 200 or more countries and cultures to get a sense of these assumptions. Researchers have found that there are five categories of assumptions that cultures share (Schein, 1984). What varies are the specific forms these assumptions take as well as their implications for management. Figure 2.4 summarizes the general nature of these assumptions and gives examples of ways they manifest in different countries.

Relationship to Nature

The first category of assumptions concerns those made about the relationship of humanity to nature. For example, in some cultures, including the United States, the assumption is that humans dominate nature and utilize it for the wealth and benefit of humankind. In other cultures, such as in Southeast Asia, the assumption is that humans and nature should coexist harmoniously. These differing assumptions lead to significantly different implications. In the United States, this assumption is an important basis for the building of dams, the mining of minerals, and the logging of trees. The implications may reach beyond these basic activities, however, to strategic planning or management practices in business. Consider how most U.S. firms view the business environment and how they strategically approach it. Is the business environment viewed as something that people must accept and with which they must try to harmonize? Or is it viewed as something that must be mastered and dominated, if possible? For example, no one would accuse Microsoft of believing that it must harmonize with the software industry environment. In fact, Microsoft is repeatedly accused of trying to form a monopoly.

Human Nature

Different cultures also make different assumptions about people. In some cultures, the assumption is that people are fundamentally industrious; in others, the assumption is that people are inherently lazy. McGregor (1960) demonstrated the relevance of these differences in his book *The Human Side of Enterprise*.

Nature of Assumptions	Specific Assumptions	Managerial Implications
Environment (assumptions about the relationship between humans and the environment)	People are meant to dominate the environment. People must coexist harmoniously with the environment.	Strategic plans should be developed to enable the firm to dominate its industry. Firms should seek positions that allow them to coexist with others.
Human Nature (assumptions about human nature)	People are generally lazy. Work is as natural as play for people.	Implement systems for monitoring behavior and establish clear punishment for undesired behavior. Provide people with opportunities and responsibilities and encourage their development.
Relationships (assumptions about how humans should relate to each other)	Individuals have certain rights and freedoms. People exist because of others and owe them an obligation.	Individual performance should be measured and rewarded. Cooperation with and contributions to the group should be evaluated and rewarded.
Activity (assumptions about the property types and targets of human activity)	People create their own destinies and must plan for the future. People should react to and enjoy whatever the present provides.	People who fail to plan should plan to fail. Planning the future only gets in the way of enjoying the present.
Truth (assumptions about the nature of truth and reality)	Truth exists objectively. Truth is what is socially accepted.	Facts and statistics are how you convince and influence people. Opinion leaders are how you influence people and decisions.

FIG. 2.4. Basic assumptions and their implications.

McGregor argued that every manager acts on a "theory" or set of assumptions about people. Theory X managers assume that "the average human being has an inherent dislike for work and will avoid it if he can" (McGregor, 1960, p. 33). Consequently, managers who accept this view believe that "people must be coerced, controlled, directed and threatened with punishment to get them to put forth adequate effort toward the achievement of organizational objectives" (McGregor, 1960, p. 34). By contrast, Theory *Y* managers assume that "the expenditure of physical and mental effort in work is as natural as play or rest" (McGregor, 1960, p. 47). Consequently, managers who accept this view believe that

> External control and the threat of punishment are not the only means for bringing about effort toward organizational objectives. Man will exercise direction and self control in the service of objectives to which he is committed. Commitment to objectives is a function of the rewards associated with their achievement. (McGregor, 1960, p. 47)

Human Relationships

This category of assumptions concerns a variety of questions. What is the right way for people to deal with each other? How much power and authority should any one person have over another? How much of an individual's orientation should be toward meeting his or her own goals versus meeting the goals of the collective society (Triandis & Bhawuk, 1997)? In a large-scale study of 40 countries, Hofstede (1980) found significant differences in the answers to these questions. Hofstede examined the degree to which people accepted power and authority differences among employees. Hofstede found, for example, that in less developed countries, local employees rarely established contact with managers brought in from other countries. This was true in the Philippines, Venezuela, and Mexico, which had the highest levels of acceptance of power differences. In other words, individuals from these and other "high power distance" cultures thought that people with higher rank should have significantly more power, influence on decisions, and so on than people of lower rank. By contrast, people in Austria, Israel, Denmark, and other "low power distance" cultures did not think that rank entitled people to significantly more power and influence; they tended to think that people from the United States, Australia, and Great Britain ranked highest for their individual orientation. To high-individualism cultures, individual freedom and rights were the most important. In contrast, people from Venezuela, Colombia, and Pakistan ranked highest for their collective orientation. In other words, in "low-individualism" or "high-collectivist" cultures, the group's interests matter more than those of the individual. The "From the Front Lines" for this chapter touches on some of the damage that occur when a manager doesn't understand the culture where he is suddenly living and working.

Human Activity

This category of assumptions concerns what the culture considers correct behavior and expectations about whether people should be active, passive, or fatalistic. In the United States, people brag about working 80 hr a week, about having no

> ### FROM THE FRONT LINES:
>
> ### "Country-to-Country Differences: Not Something to Be Taken Lightly"
>
> The following is based on the experiences of an actual employee; however, the company he worked for preferred that we not identify him or the company in any way. The case was written by a senior HR executive in a major Fortune 500 corporation.
>
> At the time Dan Broadman (not his real name) went on his first global assignment for a large investment company, he had been with the firm for more than 7 years. Although Dan had been considered a high-performing employee during most of that time, his current position required strong people-management skills and the ability to work across the organization. Dan was weak in both these areas.
>
> After a few months, upper management decided that Dan was better suited for a management position in the company's Asian office. Dan was offered the position in large part because he had lived in Asia earlier in his career.
>
> Immediately, Dan ran into significant problems. He was angry at having been taken out of his previous position and did not fully understand why he had been removed; and although he had once lived in Asia, the culture of the country he was in now was totally different from the culture in the other country. He was also very concerned about the quality of the education his children were receiving at the local secondary school. Most important, the new position required strong communication and persuasion skills.
>
> During the 1st year of the assignment, as Dan and his family struggled to adjust, relations between Dan and the local staff in Asia became more and more strained, and word started to spread to the home office that things were not going well. Although Dan was given some feedback in hopes of improving his performance, much damage had already been done. Clearly, Dan had not been prepared for the differences in the styles of interacting at home and in his new position, and he had been so overwhelmed by culture shock that he was unable to perform at required levels. Before long, Dan was repatriated.
>
> Dan accepted an interim assignment in his company's U.S. headquarters, but after 6 months, he recognized that he was not in the right role and left the company.

time for vacations or to watch TV, and about doing several things at once on their computers. They believe in such phrases as "people who fail to plan should plan to fail." In other countries' cultures, such as those in Vietnam, Yemen, and Mexico, people believe that a preoccupation with planning gets in the way of enjoying the present. In these countries, high-strung activity is not valued and may even be seen as a waste of time and energy.

Reality and Truth

People in different cultures also have different assumptions about the nature of reality and truth and about how they are verified or established. For example, the adversarial criminal justice system in the United States is based on two assumptions: truth exists, and the fire generated by opposing views will ultimately illuminate what really happened. In other cultures such as Japan, reality is more subjective and depends more on what people believe it to be. Consequently, opinion leaders or persuasive stories rather than "hard facts" are used to influence people and business decisions.

Our purpose in discussing these five categories of assumptions is to illustrate two key issues. As mentioned earlier, assumptions are the source of values and behavior. To understand the visible artifacts of culture, we must understand the invisible values and assumptions. Like the roots of a tree, fundamental assumptions, by their very nature, are not only invisible but also are generally taken for granted. Their taken-for-granted nature makes them difficult to uncover or understand because the people who hold these assumptions are not usually conscious of them and therefore cannot easily identify and explain them to people from outside the culture. People are as unaware of how their cultural assumptions affect their behavior as they are of the oxygen they breathe. Breathing and following cultural norms are natural and almost automatic processes. Furthermore, the taken-for-granted nature of cultural assumptions makes them difficult to change. Behavior may change, but underlying values and assumptions remain and are not changed either easily or quickly. For example, Japanese businessmen no longer wear kimonos, having opted for Western clothing. However, just because they made this change in behavior does not mean that conformity is no longer a fundamental value in the culture. The Japanese still value conformity, as evident in the fact that nearly all Japanese businessmen still wear the same attire: a dark suit, a white shirt, and a conservative tie. In other words, manifestations of key cultural values may change but not the actual values.

This observation brings us back to the discussion of why cross-cultural adjustment is difficult. Although every culture has both visible components (artifacts and behavior) and invisible ones (values and assumptions), the invisible components are more important because they are the source of the visible aspects of the culture. Unfortunately, because most expatriates have neither a mental road map nor a guide to a culture's traffic signals, they encounter everything from close calls to fatal cultural crashes. Consider the case of Gerald Carlson who was assigned to the Kuala Lumpur (KL) Malaysian office where he was to serve as general manager of Pittsburgh-based PENNBANK's merchant banking operation (Dennis & Stroh, 1993).

Case 4: PENNBANK Malaysia

Gerald Carlson

Gerald Carlson had worked in a wide variety of managerial positions for PENNBANK over 8 years. During that time, he and his family (wife, Susan; 10-year-old son, Johnny; 13-year-old daughter, Kit; and dog, Tipper) had moved to four different cities in the United States as part of Gerald's corporate development. His most recent position had been as vice president of corporate affairs of PENNBANK's Chicago branch. Susan had taken courses in four different colleges because of the family's moves and had finally become qualified to work as a school social worker in the local high school system.

PENNBANK Goes Global

Two years before Gerald's reassignment, PENNBANK had expanded its operations to the international arena, with a particular emphasis on establishing a presence throughout Southeast Asia. Malaysia, a strategically located nation of slightly more than 17 million people (14 million in peninsular Malaysia and the rest in Sabah and Sarawak on the island of Borneo), had been targeted as the first Asian site for PENNBANK. PENNBANK had also examined possible sites in Europe and North Africa but wanted to establish a branch in Malaysia first. Malaysia had attracted foreign investments because of its solid position in commodities—tin, palm oil, rubber, tropical hardwoods, and most recently, offshore petroleum. Thus far, PENNBANK had successfully established itself as a merchant bank in KL, the capital city of Malaysia, by providing financial support for commercial and industrial ventures. Malaysia preferred that new foreign banks enter the economy on that level. The government generally restricted the nature of a new bank's investments, ensuring that all major investments were part of a multibank coalition with other established banks in Malaysia. Malaysia placed particular emphasis on coinvestment with Bumiputera banks.

Bumiputera, *often shortened to* Bumi *or* Bumis, *literally means* "princes of the Earth" *and generally refers to indigenous Malay people as opposed to Chinese Malays or Tamil Indian Malays. The term is also used to describe those Chinese or Tamil Indian Malays who have established a three-or-more generation heritage in Malaysia. Additionally, Malays can become Bumis by converting to Islam. Although the Bumis constitute 54% to 56% of Malaysia's population, they have historically controlled only a small percentage of the Malaysian economy. Consequently, the government has formal policies to encourage the Bumis to become active in business, including in the banking community. Thus, initial start-up banks such as PENNBANK were expected to hire a disproportionate share of the Bumi population (i.e., greater than the 54% to 56% present in the population) and to assist in financing Bumi-owned or controlled projects. This requirement has* constituted a kind of affirmative action policy on behalf of the majority group. In addition, the government has generally prohibited start-up foreign banks from establishing full-service banking positions in Malaysia for several years. Expansion of PENNBANK's ability to lend and eventually take a major position in financing large-scale commercial and industrial projects depended on the development of a goodwill relationship with the Malaysian government, the Malaysian banking community, and the Bumiputras.

As the newly assigned general manager of its 2-year-old KL main branch (PENNBANK–KL), Gerald had limited experience in Southeast Asia. He had been sent to KL on several extensive trips during the period that PENNBANK was establishing itself in the local economy. During those trips, Gerald worked well with the local people. For Gerald, this assignment was a major promotion in status, responsibility, compensation, and benefits. It also represented a significant challenge and change in duties with regard to locale, type of banking he was involved in, and cultural diversity.

PENNBANK–KL'S Need for a New GM

The previous and first general manager of PENNBANK–KL, a single man, had just been removed from the assignment after he was charged, by a person unknown to him, with violating Malaysia's "close proximity" law regarding Muslim women. This religious law states that if a citizen of good repute reports than any man is seen in close proximity to an unmarried woman, such as holding hands in a park or walking together on the beach in the evening, the man must immediately marry the woman or serve a jail term. Although not usually enforced in KL but often enforced in some of the more traditional states of Malaysia such as Terengganu and Kelantan, the close proximity law has been used as a device to "control" foreign competitors. That is, competitors have been known to seek out opportunities to observe improper behavior and report it to the police. Amid reports that he had supposedly breached the law, PENNBANK's previous general manager had been hurried out of the country. The charges were subsequently dropped; however, PENNBANK received considerable adverse publicity in the Malaysian newspapers. The acting assistant general manager, Abd Mahmoud bin Malek, a local Bumiputera, attempted to fill the vacancy, but he had only limited previous experience with PENNBANK. He did not do well, and no other local replacement with sufficient experience and ability was available to fill the job.

In the meantime, some of PENNBANK–KL's more fundamentalist Bumi employees continued to express outrage over the past general manager's behavior and demanded that PENNBANK hire a general manager from the local community. PENNBANK's executives in Pittsburgh believed that the branch was not yet well-enough established in the Malaysian economy to justify hiring a host-country national. In fact, executives in the home office thought that they had no choice but to bring in another general manager immediately.

Gerald's Big Chance

On his selection as the new GM, Gerald was given a fast-track preparation course. He spent 1 day with several other PENNBANK officers who had previously visited KL, 2 days with people at the Malaysian Industrial Development Authority's Chicago office, and a day with staff in PENNBANK's human resources department in Pittsburgh with whom he discussed temporary living arrangements, flight schedules, allowances, compensation, benefits, deferred income, Malaysian taxes, and so on. Gerald was also given a copy of a 1970 book, The Malay Dilemma, by Mahathir bin Mohamad, the prime minister of Malaysia, for background information on the economy of the region. Gerald then had a week to pass on his job responsibilities to his successor at PENNBANK's Chicago branch. Neither Gerald nor his family received any other cultural, historical, language, problem-solving, or skill-building preparation.

The Family

Fortunately for the children, the move occurred in the summer, so school was not disrupted. However, when the children arrived in KL, it was the middle of the school year, which starts in June.

When the Carlsons arrived in KL after a 32-hr trip via Tokyo and Hong Kong, they were booked into a luxury hotel until they could find permanent accommodations. It took Susan 6 weeks to find a house, clear the sale, and secure permits for electricity, water, telephone, and television (all separate government monopolies). When they moved in to their new home, the Carlsons were assigned a live-in maid/cook. Because of Gerald's new status in the KL expatriate community, Susan was expected to be a member of the American Club and of several country clubs, and the couple was frequently invited to U.S. embassy receptions.

Life changed very radically for Susan and the children. Susan quickly discovered she could not practice her new profession of school social worker because Malaysia (like the United States) issues work permits to only the husband or his wife if they are foreign aliens. The family also discovered it could not get its dog, Tipper, out of quarantine. Malaysia has strict rules against importing dogs because they are considered unclean in the Muslim faith. Tipper was eventually "put away," much to the consternation and trauma of the entire family.

In addition, instead of relishing the freedom of summer vacation, the children found themselves in a new private school dealing with a new curriculum and a new language. The other students in the school were mostly embassy offspring and came from virtually every country with an embassy in KL, making for an interesting and varied assortment of children. The school had a decidedly British orientation, so Johnny and Kit were not happy with their course of studies or manner of instruction. They began to object to the mandatory course in Bahasa Malaysia because, as they insisted, "Everyone speaks English anyway, and besides, we're Americans, not Malay."

Problems at PENNBANK-KL

In the meantime, Gerald was experiencing many problems, leading to increasing feelings of frustration. He quickly noted that the acting general manager had hired several of his Bumi relatives despite PENNBANK'S strict policy against nepotism. Gerald was also disturbed that most of the locals were going to his assistant, Mahmoud, with their problems instead of directly to him. Gerald was concerned as well with the obvious difficulties the various ethnic/racial/religious groups seemed to have getting along each other. The Bumiputeras claimed they were entitled to a higher percentage of the jobs. The Chinese Malays seemed to go about their business well but were unwilling to provide assistance to the Bumi. The few Tamil (Indian) Malays seemed to have a resigned attitude about their promotion potential and defined their jobs in very limited ways. Gerald also noticed that supervisors were unwilling to accept responsibility, and his efforts at delegating were to no avail.

On the positive side, new accounts, which had virtually dried up before Gerald had arrived, were making a steady but slow recovery. Gerald had personally participated in packaging several small project loans with Malaysian Banking Berhad. In short, the job was tough but rewarding.

The Ultimate Ultimatum

At Susan's insistence, at the end of their 3rd month in Malaysia, Gerald and Susan discussed returning to the States. The children seemed to be experiencing negative effects from the move; in particular, Johnny would not leave the house, and Kit had become friends with an unruly group of embassy children. Gerald and Susan were concerned about the long-term effects of the assignment on the children. Gerald acknowledged that the position was not as exciting as they had been led to believe, although he was making progress. He was becoming increasingly frustrated with the energy level required to do what he had determined was a mediocre job at best, but he still wanted to stick with the job and finish the assignment. Susan missed her career more than she had anticipated. Susan and the children wanted to go home, and they made this clear in no uncertain terms. Family pressures had built up, and after several heated discussions, Susan and Gerald finally agreed that he should tell his boss that the new position had not worked out for either him or his family.

The next day, Gerald sent a fax to his boss in the United States. The tone of the document was quite negative. Gerald reminded his boss that he had promised there would always be a job for Gerald in the United States if he agreed to go to Malaysia and help out PENNBANK. Gerald reported the following problems and frustrations with the job:

1. The Malaysian government's limitations on banking.

2. Frustrations with a no-nepotism rule in a country where nepotism was seen as a normal way of life, making it impossible for Gerald to dismiss some of the deadwood at PENNBANK–KL.
3. The constant infighting between himself and Mahmoud coupled with Mahmoud's popularity among the employees. How could he ever gain the legitimate power needed to run the bank as needed?
4. The incessant struggles among the various ethnic/religious groups.

Gerald did not mention the family's personal reasons for wanting to return home, but there were many:

- Susan was tired of having the maid follow her around all day and that she was always giving sweets to the children.
- Susan did not want to spend another hour with the vapid alcoholics at the American Club.
- Susan was sick and tired of being in a country where she couldn't use her skills and work outside the home.
- There were no age-appropriate friends for Kit in the area.
- Kit and Johnny were both frustrated by the requirement that everyone had to study the Bahasa Malaysia language in school, even though everyone spoke English.
- Kit and Johnny were further frustrated by the fact that mostly Bahasa Malaysian films were shown on television. The American and English films were usually played after their bedtime.
- Johnny was failing his course in the history of Victorian England and couldn't understand why he couldn't study American history instead.
- The home telephone had stopped working for the third time that week.

THE PROCESS OF CROSS-CULTURAL ADJUSTMENT

The episodes depicted in the preceding case are common. Many global managers and their families experience adjustment difficulties similar to those of the Carlsons. Most of the time, the spouse, quite often the wife, has to give up a job, a house, friends, and family to accompany her husband on a foreign assignment. Consequently, she may have even greater difficulty than her husband adapting to the foreign culture. In chapter 5 (this volume), we discuss some of the factors that affect a spouse or partner's adjustment as well as that of the children. For now, we concentrate on the dynamics of cross-cultural adjustment. This process requires serious consideration, for it drastically disrupts the individual's routines and dramatically affects the individual's ego and self-image.

Routines

Almost no one wants total uncertainty in life. In fact, most people want a reasonably high degree of certainty and predictability. That is primarily why people establish routines. The global success of McDonald's is testimony to the general human need for a certain level of predictability. When we go into a McDonald's, there are a variety of items from which to choose, but we know how a Big Mac is going to taste before we order it. A Big Mac is a Big Mac. People like not only the taste of the hamburger but also the fact that they know what to expect when they enter a McDonald's, whether it's around the corner from home or on the other side of the globe.

People's routines affect all aspects of their lives, from the mundane to the critical. For example, people establish routines for waking up in the morning: shut off the alarm, get up, take a shower, get dressed, eat, run out the door. People also establish more serious routines regarding initiating and developing relationships, dealing with conflicts, and their expectations about relationships.

Routines and the certainty they provide create a kind of psychological economy in our lives. Routines enable us to process far more details than we could ever attend to if every activity required the same conscious thought and attention. We establish routines because we cannot consciously process an infinite number of issues simultaneously. Because we know that once we get up we will take a shower, we do not have to devote a lot of time and energy to thinking about and processing those decisions. When a routine is disrupted, however, more time and energy must be devoted to processing even mundane activities. Mental time and energy are limited; therefore, the disruption of routines decreases mental time and energy available to devote to other issues.

Not everything in an unfamiliar cultural environment is equally difficult to adjust to. Whether the adjustment is relatively easy or extremely difficult is a function of three dimensions: the scope, magnitude, and criticality of the disruptions in everyday routines.

Scope. The greater the number (or scope) of disrupted routines, the more difficult the process of dealing with the disruptions and the greater the frustration, anger, and anxiety that are likely to follow. Having your morning-shower routine disrupted is one thing, but having eating, sleeping, commuting, and working routines disrupted is quite another. Similarly, having to give up a handshake for a bow when greeting someone may be inconvenient, but having to alter your entire management style, including how you delegate authority, make decisions, influence people, plan and organize the workday, and motivate subordinates, can be truly upsetting.

Magnitude. Disruption is experienced along a continuum from slight alteration to total destruction. The greater the magnitude of the disruption, the greater the time and energy required to deal with it and the greater the frustration, anger, and anxiety you are likely to experience. If taking a shower was always the first task of the day, then having to take a bath would be somewhat less of a disruption than not being able to do either without going to a public bathhouse. It may be somewhat irritating to have to switch from cash bonuses to days off as motivation incentives, but it can be totally frustrating to have the option to use incentives completely removed and in the hands of a labor ministry.

Criticality. Some routines are a central part of daily life; others are trivial. The greater the criticality of the routine that is disrupted, the greater the time and energy required to deal with it and the greater the frustration, anger, and anxiety that are likely to follow. Most people would probably say that not having a reserved parking spot is less frustrating than having to change from a tell-and-sell style of leadership to one that entails consultation and consensus decision making.

Culture Shock

So far, we have noted that (a) living and working in new and unfamiliar cultural environments disrupts routines; and (b) the more routines are disrupted, the more severely any given routine is altered, and the more critical the disrupted routines are, the greater the time and mental energy required and the greater the frustration, anger, and anxiety that result. Although these statements are true, they don't explain the severity of the symptoms that people in an unfamiliar culture often suffer. In combination, these psychological and emotional responses are called *culture shock*. The symptoms include frustration, anxiety, anger, and depression. However, unlike the level of these emotions that are experienced when one or two routines are disrupted, the level of depression and even anger people experience with culture shock is often very severe.

To fully understand culture shock, it's important to understand the entire process of expatriation from the point of departure for the international assignment. Typically, the first few weeks or month of an international assignment are a "honeymoon" period. During this phase, the assignee and family may be violating cultural rules; however, they do not realize they are doing so, and more important, they are unaware of negative feedback. They don't know enough about the culture to know what cultural mistakes they are making, so they miss negative signals from locals.

An even more powerful explanation exists for this honeymoon phase. This explanation lies in something that for most of us is quite fragile—our egos. Most of us have positive self-images that we want to protect. Few of us want to look like fools if we can avoid it. Consequently, early in an international assignment, we tend to ignore or downplay any negative signals we pick up. However, after a while, the sheer number of negative signals becomes overwhelming, and we can't deny or ignore them any longer. Culture shock sets in (Janssens, 1995).

If it is true that for every action there is an equal and opposite reaction, is the reverse also true? For every reaction, is there an equal and opposite cause? One would expect this to be true in the case of culture shock.

It may not seem obvious, but routines are a reflection of the self. In a sense, a routine demonstrates a level of proficiency that is usually taken for granted. Living in a foreign culture challenges these basic proficiencies, making you extremely aware of areas of weakness and strength. In fact, the more proficiency is taken for granted, the more severe the reaction to its loss is likely to be.

A closer examination of Gerald Carlson's case should make the close interconnection between these components of cross-cultural adjustment much clearer. A few months into his assignment in Kuala Lumpur, Gerald found himself frequently so angry that he had to struggle not to vent his anger by striking someone. Sometimes, he was so depressed that he could find very little reason to get out of bed in the morning. The extent of these emotions was not visible on the outside, and Gerald worked hard to keep it that way. In Gerald's case and in other severe cases like his, there has to be more than a disruption of routines to explain why Gerald and his family would pack up and head home or why one in five Americans leaves a foreign assignment prematurely.

For Gerald—someone who had been to countless important dinners with clients and who had developed an impressive ability to deal with these situations smoothly—the interpersonal problems with his assistant Mahmoud and his employees was a significant blow to his self-confidence and ego. These assaults on his self-esteem and not just the disruption of routines led to Gerald experiencing culture shock in the extreme.

The more basic the routine that is disrupted because of cultural ineptitude, the more severe the blow to the ego and the more severe the resulting culture shock. For example, getting around in the city in which one lives is a skill that is often taken for granted. For Susan Carlson, driving around Pittsburgh, even in heavy traffic, was something she did without even thinking about it. Being unable to find her way from her house to a friend's was a severe shock to her self-image as an independent, capable person. Foreign assignments involve a steady stream of such incidents, from the simple to the complex, that challenge self-image. Expatriates and family members are constantly confronted with situations that send certain messages: "You don't understand this," "You can't do that," "Even 6-year-olds in this country know that," and "You're an idiot."

As the number and magnitude of demoralizing incidents build over time, people get worn down and can no longer ignore them. Although the specific symptoms vary among individuals and even within individuals from week to week, anger and frustration are common. Anxiety and depression are also common as the person's positive self-image gets battered and confidence crumbles. Quite often the inherent mechanisms by which people defend and maintain their egos cause them to direct their frustration toward others. This is a primary reason that people blame others, notably the locals. In any gathering spot for Americans, one is likely to hear conversations peppered with statements like these:

- I can't believe how stupid these locals are. Their street addresses make absolutely no sense.
- These people think they're so superior to the rest of us. I think locals really resent people from other countries.
- The locals are just plain lazy. It's impossible to motivate them, and they feel no loyalty to the company.
- This whole thing is my spouse's fault. He has no appreciation for what I'm going through. He has his comfortable little cocoon at the office.

Unfortunately, many people never recover from culture shock. Some return home early but not all. Of those who never recover, many stay for the duration of their assignments, usually fearing the consequences of returning early or hoping their situation will improve over time.

Most of those who stay eventually work through their culture shock and gradually adjust to living and working in the new culture. The pain of making mistakes is the primary source of the feelings of inadequacy and depression, but it can also be a source of adjustment. Once a cultural mistake is made and, more important, recognized, it is not likely to be repeated. Gradually, by making mistakes, recognizing them, and observing how others in the culture behave, people learn what to do and what not to do (Ward & Kennedy, 1993).

SUMMARY

In this chapter, we have elaborated on several general statements about cross-cultural adjustment. People establish routines to obtain predictability in life and to achieve psychological economy. Routines also provide an important means of preserving and maintaining ego and self-image. Living and working in new cultures disrupts established routines. The more routines are disrupted, the more severely a given routine is altered; the more a disrupted routine is critical, the greater the time and mental energy required to cope and the greater the frustration, anger, and anxiety associated with the culture shock that occurs. Most important, however, disrupted routines are generally accompanied by situations that challenge an individual's confidence, ego, and self-esteem. Threats to these sensitive areas cause the strongest reactions associated with culture shock—depression and even hatred. In principle, circumstances that increase disruption and uncertainty tend to inhibit cross-cultural adjustment, whereas circumstances that reduce disruption and uncertainty tend to facilitate cross-cultural adjustment. Having outlined culture shock and adjustment, we can now focus on the specific processes of effectively selecting and training managers for international assignments.

PART II

BEFORE THE ASSIGNMENT

CHAPTER
3

Selecting: Identifying Candidates With Global Leadership Potential

Case 5: Mathison

Doug Sweedlow, director of Mathison's Information Technology (IT) department, was starting to feel frantic. He had promised to have somebody in Brazil in a couple of weeks and still no one had been chosen for the job.

Pressure was mounting from corporate headquarters in Los Angeles. "We need to get somebody there fast, Doug. The plant is having no end of problems!"

Because of delays in the Brazil plant, Mathison had missed several critical deadlines for rolling out software. These delays had been costly because the group there was an essential component of Mathison's software-development team. When Mathison got behind, production got delayed in other plants as well. Corporate headquarters wanted an immediate change in the upper rungs of the Brazil operation to ensure that production got back on schedule and to set up safeguards to prevent the plant from getting behind in the future.

Doug spent the rest of the morning reflecting on who had the best skills in IT to effectively solve the problems in Brazil. After mentally reviewing the best programmers in his division, he came up with a short list of three good candidates for the job.

As Doug reviewed the candidates, one name kept jumping out at him: Mike Tompkins. Mike was one of the best programmers Doug had in the United States. He also fully understood Mathison's approach to software development, and most relevant to Doug, Mike had done a similar job in Atlanta when a group there had problems. Convincing himself more and more that Mike was the man to send to Brasilia, Doug set up a time to interview him.

Two days later, Mike flew to Doug's office in Los Angeles. During the interview, Doug explained that the job in Brazil would not be easy. The Brasilia group had been consistently behind, and its new manager would have to turn the situation around. Mike assured Doug that he was well prepared to tackle the problems; after all, he had succeeded in Atlanta. Doug emphasized that on one hand, the corporate office would be eyeing him like a hawk; the company had already lost a sizable amount of income by getting behind and was eager to get the Brazil team up to corporate standards. On the other hand, Mike could expect a lot of perks and perhaps a promotion if he succeeded in this high-visibility assignment.

As the interview was wrapping up, Mike told Doug that he was convinced that his family would be very excited about the opportunity to live in Brazil for a few years and that the assignment obviously represented a fantastic opportunity for him. He couldn't wait to get home to discuss the offer with his wife and children.

CROSS-CULTURAL VERSUS NARROWLY FOCUSED APPROACHES TO SELECTION

Mathison's response to its software-development problems in Brazil was typical of the approach many organizations use to select candidates for global assignments. The company's approach also reflects why this approach so often fails (Black & Gregersen, 1999; Harris & Brewster, 1999).

Basically, a crisis had arisen in a foreign operation and Mathison had little time to assess the problem strategically or systematically. Consequently, the selection process was motivated by a strong desire to solve a specific problem (delays in software development) and to do so quickly. As a result, the focus of the selection process was to put the fire out. This resulted in an obsessive-like focus on finding someone with strong IT and managerial qualifications that would presumably be able to solve what management saw as the short-term problem in Brazil.

By reviewing some of the conscious and unconscious decisions Doug made in offering Mike Tompkins the position in Brazil, we can begin to see how a company's selection process contributes to the high failure rates for global assignments. First of all, Doug overlooked the ways in which Mathison's human resource department might have been able to help him both identify potential candidates for the Brazil position and evaluate them. This would have expanded the list of candidates to include people Doug did not know. As it was, Doug considered only people with whom he was very familiar.

Second, Doug failed to consider the ability of the candidates and their families to adjust to and function effectively in a new cultural environment. A technically oriented selection process like the one Doug used can easily result in costly premature returns or ineffective performance during an assignment—just what Doug wanted to avoid. In retrospect, Doug should have used every available resource, from people in the human resource department to other managers, in generating a complete set of candidates.

Premature returns or ineffective performance are often direct results of firms selecting technically qualified candidates who lack the cross-cultural communication or adjustment skills to perform well in a different culture. The case of a senior executive with a U.S. carmaker illustrates the severity of the damage that can occur when a person lacking cross-cultural skills is sent on an international assignment. In this case, the assignment was in Korea. The senior executive had been picked in large part because he was especially adept at negotiating contracts. As the company soon discovered, his confrontational approach, which had worked well with American steel suppliers, was very offensive to the consensus-minded Koreans. The man so offended his suppliers that before long they would not even talk to him. Not only was he called back to the home office, but it took his replacement a year to undo all the damage (Black & Gregersen, 1999).

Another reason international assignments often fail is reflected in the practice of not carefully considering the entire family situation—matters Doug apparently overlooked completely. This often results in the family wanting to return home early or at the very least suffering emotionally when someone in the family encounters severe cross-cultural difficulties. As one U.S. human resource executive told us, "For 24 years I have seen expatriate families come and go; many fail because the family can't adjust." Research supports this executive's comment. According to one recent survey, a partner's dissatisfaction accounts for 27% of the early returns from global assignments. Family concerns account for another 26% (Windham International et al., 2002).

Some failures could be avoided, however, if multinational firms would redesign their candidate selection practices.

COMMON SELECTION PRACTICES

Doug felt pressured by corporate headquarters to quickly select a candidate who would improve the performance in the Brazil operation. Doug knew that his success in Mathison would seriously depend on the success of his selection for this assignment. Minimizing any risk of failure was critical because failure would ultimately reflect on Doug's performance as well as that of the person he sent to Brazil. This approach was not inherently wrong; everyone wants to succeed. What was wrong was the inadequacy of the process Doug used to select the person for this global assignment. Unfortunately, Doug's attempt to minimize the risk of failure probably maximized that risk.

In choosing Mike for the international assignment, Doug appeared to have relied too much on what Harris and Brewster (1999) called the "coffee-machine system"—that is, selecting people based on personal recommendations while getting a cup of coffee. As a result, the pool of candidates is limited to employees whom managers know, eliminating many potentially better candidates with whom these managers are less familiar.

Failure in a global assignment (poor performance or premature return) most often occurs because international assignees and their families are unable to adjust, not because the employee lacks technical or professional abilities. In fact, the successful completion of a global assignment is often linked directly to the expatriate's and the spouse or partner's cross-cultural adjustment. In other words, companies all too often send people on international positions who are capable but "culturally illiterate" (Black & Gregersen, 1999). This is borne out by alarming statistics. In one study, almost 100% of the respondents said that when choosing candidates for international assignments, skills were the "most important" or a "very important" selection criterion (Black & Gregersen, 1999). That technical skills receive such disproportionate attention may be a function of the limited role that human resource departments have traditionally played in the selection process (Halcrow, 1999; Stroh & Caligiuri, 1998).

In a very important study of global assignment selection processes, Miller (1973) examined the activities that managers engaged in before making selection decisions. Miller found that when decision makers did not quickly identify candidates with high qualifications (technical, job-related competence), they were more likely to carefully define the range of skills required, to determine more precisely how to measure performance during the assignment, to search more aggressively for potential candidates throughout domestic and international divisions (by seeking references from fellow managers or reviewing personnel files), and to request more assistance from human resource departments. Essentially, a paradox is inherent in the selection process of many U.S. firms. When no candidates with high technical and job-related qualifications are immediately identified, line managers extend the search throughout the corporation, locating individuals with superior cross-cultural adaptation and communication skills as well as strong professional qualifications; but when one or several technically qualified candidates quickly come to the attention of decision makers, they terminate their search and often overlook candidates with equally strong technical qualifications and superior cross-cultural skills.

COMMON SELECTION PRACTICES USED OUTSIDE THE UNITED STATES

Before noting similarities and differences in cross-cultural selection practices, it is important to note that all selection practices should begin with an understanding of local and national laws and regulations regarding hiring, promoting, and relocating employees. For example, in the United States, one must have a basic understanding of Equal Employment Opportunity Commission laws that govern this process. In Europe, for example, one must understand the principles of the Data

Protection Act that in some instances governs whether personnel data can be transferred across borders via electronic transmission. Without this knowledge, those involved with an international assignee selection might soon find they have violated local and national laws that may effect the processes they use for selection. Thus, we first encourage those involved with the selection of international assignees to become familiar with country-specific national and local laws relevant to the selection process.

Firms in Japan, Western Europe, and Scandinavia generally use selection processes similar to those used by firms in the United States. Tung's (1988) study of the selection process in the United States, Japan, and Western Europe found that "managerial talent" was one of the top selection criteria in all three geographic areas for the selection of CEOs of foreign operations. In Scandinavia, professional qualifications are also the predominant criteria line managers use in selecting personnel for global assignments (Björkman & Gertsen, 1992; Gertsen, 1989; Kainulainen, 1990). Essentially, in selecting global managers, multinationals throughout the world tend to focus on finding individuals who exhibit the highest professional or managerial qualifications.

However, although similarities exist between the selection processes in the United States, Japan, Scandinavia, and Western Europe, there are differences among these processes as well. For example, interviews are almost always conducted in the United States (99%) and Western Europe (100%) but are held less frequently in Japan (71%) and Scandinavia (75%). These differences are even more apparent when we consider how frequently spouses or partners are interviewed before decisions about global assignments are made. In the United States, spouses are sometimes interviewed or briefed before an employee is offered a global assignment (52%), less often in Western Europe (41%), rarely in Scandinavia (18%), and never in Japan (Björkman & Gertsen, 1992; Katz & Seifer, 1996; Nicholson & Ayako, 1993). These differences, however, do not necessarily result in better or worse assessments of candidates' and partners' abilities to adjust. The differences may stem from unique cultural factors.

In Japan, for example, the family is not an issue in the selection process; if a man is advised to make an international transfer, the effect of the assignment on the family is not considered relevant because Japanese decision makers believe that a wife will not really be able to influence her husband's decision. Even if a Japanese wife was not willing to move overseas, her husband would still be bound to the firm and would have to take the assignment (White, 1988). It is important to note, however, that the cultural homogeneity and paternalistic practices of Japanese business do provide mechanisms for identifying potentially difficult situations involving the spouse and the family.

Cultural reasons may also explain why so few Scandinavian firms evaluate spouses. In Scandinavia, strong respect for personal privacy is exhibited along with an implicit expectation that home life will not be subject to formal, organizational evaluation. As in Japan, however, the smallness of the Scandinavian countries and the relative homogeneity of each one provide opportunities for firms to learn about potentially difficult family situations without relying on extensive formal evaluations.

There are also differences among countries in the degree to which personality or skill tests are used as selection methods. Line managers and human resource professionals agree that candidates' ability to communicate with and relate to people across cultures is important to the success of international assignments, but very few firms actually test these skills formally before offering someone a job abroad (Deller, 1997; Ones & Viswesvaran, 1997; Zeira & Banai, 1985). Specifically, 24% of Scandinavian and 21% of Western European firms rely on formal testing mechanisms to evaluate candidates' abilities to relate to people across cultures. In contrast, only 5% of U.S. and virtually no Japanese firms administer tests to assess such skills (Björkman & Gertsen, 1992).

U.S. multinationals rely especially on the suggestions of line managers and human resource staffers who often have little knowledge of the business and social culture into which the potential expatriate will be placed. Further, as little as 20% of companies bother to assess whether candidates have the well-known core personality traits and competencies that enable someone to succeed in a foreign culture (Ioannou, 1995).

Collectively, these findings indicate that European and especially Scandinavian firms may be slightly more strategic and systematic in selecting international personnel because they utilize a wider variety of evaluation methods and pay more attention to cross-cultural skills (in addition to technical qualifications). Although it is difficult to draw definite conclusions, the greater emphasis in Scandinavian and European firms on assessing these skills during the selection process seems to relate positively to expatriate adjustment and performance.

IMPORTANCE OF STRATEGY TO THE SELECTION PROCESS

To acquire or maintain a competitive position in the global marketplace, a firm must seek the highest possible return on its investments in international assignments. The first step is to integrate strategy into the selection process of global employees. As the firm increases its global reach and moves through the various stages of globalization, assessing this process becomes increasingly important. For example, a firm moving from the export stage to the coordinated multinational stage of globalization must plan strategically for the future because the need for qualified international assignees will be much greater as the firm moves out of the export stage. Without such organizational foresight, the firm will undoubtedly reach a future stage of globalization only to discover it has a shortage of qualified personnel to send on international assignments.

The strategic importance of selecting the right person for a job is further reinforced by examining international joint ventures. According to some estimates, international joint ventures fail at a rate of 50% to 75%, and these failures occur at least in part because expatriate managers are unable to adjust to local business practices (Goodman, 2003). Furthermore, firms that fail to approach the selection process from a strategic perspective are likely to be doomed to send people on global assignments who have strong technical qualifications but are lacking in cross-cultural skills.

In chapter 1 (this volume), we discussed three central strategic roles of international assignments: leadership development, coordination and control, and information and technology exchange. In the Mathison case discussed at the beginning of this chapter, Mike Tompkins's assignment to Brazil could have served the important function of enhancing information and technology exchange; however, Doug Sweedlow's hasty selection of Mike virtually ensured that this objective would not be systematically accomplished. For example, unique software-development issues impeded the group in Brazil from meeting its deadlines. As a major step in solving these problems, Mike could have consulted with IT experts in Brazil and forwarded this information to other divisions of Mathison. However, because he lacked cross-cultural skills, he never took such action, further delaying the resolution of the Brazil group's problems.

Identifying candidates with the potential to develop global leadership skills and outlooks also would have furthered Mathison's strategic goal of improving leadership development. The selection of a candidate with both technical and cross-cultural skills would have resulted in the development of a general manager who could have assumed important executive positions in Mathison's worldwide operations. Although Mike was technically capable, he did not possess the general management potential needed for a position as a senior executive in the firm. Mathison's mistake—one many firms make—was that it did not identify and let Doug and Mike know the long-term strategic goals of the global assignment, thereby sacrificing long-term objectives for short-term results. Had the company had a more systematic selection process, it could have achieved both its long-term and its short-term objectives.

When selecting individuals for international assignments, management must remember that advancement to the next stage of globalization is accomplished in unique cross-cultural contexts. Accordingly, specific factors need to be considered in the selection process.

FACTORS TO CONSIDER IN SELECTING INTERNATIONAL ASSIGNEES

Practicing managers and international researchers have developed long lists of critical factors to consider when selecting candidates for global assignments. In general, these lists can be divided into expatriate- and spouse- or partner-related categories. Most important, however, decision makers need to remember that the fundamental purpose of the selection process is to choose individuals who will stay for the duration of their global assignments and accomplish the strategic and tactical goals they set out to accomplish.

Differences by geographic regions seem to exist with respect to some of the criteria used to select candidates for global assignments. For example, U.S. firms place less emphasis on language skills than do European and Japanese companies. Also, in contrast to the consistent focus on technical skills in U.S. companies, European firms rely more on spouse- and family-related considerations than do U.S., Scandinavian, or Japanese companies (Nicholson & Ayako, 1993; Tung, 1988).

Strategic Factors

As we said earlier in this chapter, before selecting candidates to send overseas, it is very important to assess the critical strategic objectives of the international assignment. To achieve all of these objectives may require that the person on the international assignment possess several skills as well as a range of experiences and contacts. For example, if the primary purposes of the assignment are to improve the control function between headquarters and the subsidiary and to increase the coordination function between subsidiaries, then the candidate should have broad experience in the firm including a wide array of contacts throughout the company. Another strategic purpose of an international assignment may be to exchange critical information between the foreign operation and headquarters. This exchange may require the movement of information not only from headquarters to a subsidiary but also from the subsidiary back to headquarters. To perform this function, a candidate must have not only the necessary information from headquarters but must also possess excellent cross-cultural communication skills because the information must be conveyed to people in the subsidiary, and important information acquired from the subsidiary must be transmitted back to headquarters. If the strategic purpose of an assignment is management or executive development, then the candidate's experience within the firm and his or her advancement potential should be important criteria in the selection process.

Of course, strategic functions are not mutually exclusive, and the selection criteria relevant to performing one function (e.g., coordination) are often relevant to performing another (e.g., information exchange). Decision makers must pay attention first to defining the strategic objectives of the global assignment and then to carefully assessing the skills, knowledge, and experience required to accomplish those objectives.

Professional Skills

Whether the job assignment is for a CEO, a functional department head, or a technical specialist, professional skills (either managerial or technical) are essential. These skills generally include direct knowledge of the job and a grasp of the specific problems to be solved. For example, in the Mathison case, Mike Tompkins needed strong knowledge of information technology as well as managerial ability. However, as Mathison found out, although it was necessary to send someone with a strong IT experience, that alone did not guarantee success or maximize the firm's investment in the global assignment.

Conflict-Resolution Skills

In domestic as well as international managerial positions, how individuals resolve conflicts can have a significant impact on the success of an international assignment. Interpersonal conflict is a primary source of stress during these assignments (Clarke & Hammer, 1995). More important, the ways in which individuals resolve conflict can have a significant impact on their effectiveness in the assignment. For example, studies of Japanese and Canadian managers have found that the inability to deal collaboratively with cross-cultural interpersonal conflicts is related to failure in adjusting to new cultures (Abe & Wiseman, 1983; Black, 1990). The collaborative approach to conflict resolution is important because it helps individuals focus on understanding the viewpoints of the people in the other culture instead of forcing the other people to see situations the expatriate's way.

Leadership Skills

The leadership styles of expatriate managers can also have a significant impact on their effectiveness during global assignments. Research suggests that a management style that is defined as "feminine" in the United States works well in many parts of the world because this style is more the norm (Adler, 2002; Stroh, Varma, & Valy-Durbin, 2000). In general, research indicates that in other cultures, high-involvement management, which focuses not only on accomplishing tasks but also on paying attention to people, is generally superior to other managerial styles. Management research has also found that trusting fellow employees and involving them in the decision-making process results in better overall decisions, greater acceptance of decisions, and increased satisfaction in domestic and international management situations (Negandi, Eshghi, & Yuen, 1985).

Communication Skills

It should almost go without saying that the ability to communicate is crucial to success in global assignments. Most strategic functions of global assignments require individuals to communicate effectively in other cultures. Yet companies often minimize the ability to communicate when choosing candidates for global assignments.

Research has found several important dimensions of the cross-cultural communication process relevant to expatriate managers. One survey (Clarke & Hammer, 1995) found that interpersonal communication skills are one of the most important factors in influencing the success of international assignments. Without some level of proficiency in the host-country language, it is very difficult to communicate genuinely with host-country nationals. Language proficiency is a tremendous advantage to operating in a foreign land. As one American expatriate explained, "The 'key' to understanding the host country is the language. I cannot possibly understand why companies do not provide more language training to accomplish this!"

We have also found that willingness to communicate is critical to effective adjustment during a global assignment (Black, 1990; Clarke & Hammer, 1995). Although this characteristic may seem obvious, many expatriates are simply unwilling to try to communicate genuinely with host-country nationals; they rely on subordinates and translators to communicate the "necessary information" instead of engaging in significant two-way conversations. This unwillingness to communicate can ultimately derail the achievement of the strategic objectives of the assignment because coordinating, controlling, and transferring information is that much more difficult.

Willingness to communicate is also relevant to how well partners adjust to living overseas because they often have to work hard to initiate and develop social relationships and must try to communicate with others even when others may not want to communicate with them (Pellico & Stroh, 1997). For example, one American spouse told us

> During both my global assignments, I have not once received a warm welcome or strong social support from other bank wives. I knew I would have to build my own life overseas, but I expected the first steps to be taken by others in England. My advice to future expatriate spouses? Be prepared and willing to develop contacts and friendships from day one.

Social Skills

Being open to forming new relationships and able to get along well with others are valuable traits for anyone—employee or partner—moving abroad. Having strong social or relational skills can be extremely helpful when a manager wants to develop social relationships with locals who are often able to provide critical work- and nonwork-related information and feedback on how the manager is doing. Researchers are still assessing the value of using personality characteristics to predict success in a global assignment, but it appears that managers who are extroverted and agreeable as well as emotionally stable and conscientious are more likely to complete their assignments and adjust well to new cultures than managers who are less outgoing and less able to form close social relationships (Caligiuri, 2000; Culpan & Wright, 2002; Stroh & Caligiuri, 1998).

Ethnocentricity

How we interpret what is going on around us can have a significant impact on our adjustment to a foreign assignment. We often misinterpret and criticize the behavior of people when we cross boundaries. For example, a Japanese manager negotiating with an older manager from Finland may think that the sound created by the Finn sucking in air through his mouth means that the Finn is responding negatively to the deal, but the Finn may actually be communicating agreement by making the same sound that indicates disagreement in Japan. If an American were involved, he or she might wonder if the Japanese and Finnish people have breathing problems. In all likelihood, the Japanese manager is less likely to misinterpret this behavior because virtually all Japanese multinationals insist that their multinational managers have skills in personnel relations. U.S. managers are much less likely to have these skills, which are much less valued in U.S. corporations.

Using our own rules often leads us to misinterpret behavior in other cultures. Accordingly, those international assignees who are less judgmental and less likely to criticize behavior in the new culture have a much easier time adjusting to the new environment (Black, 1990; Caligiuri, 2000). Moreover, those individuals who are less rigid in their evaluations of the "rightness" and "wrongness" of others' behavior are more likely to succeed. Individuals who see their way as the only way are referred to as *ethnocentric*. The significance of this characteristic has been reinforced again and again: International assignees and their partners from around the world report that people on global assignments must be flexible and open-minded.

Flexibility

Another important characteristic to look for in potential international assignees is openness to new experiences (Caligiuri, 2000). These might include new food, new sports, new forms of recreation, or new ways of traveling. For example, when Americans visit Japan, are they willing to eat sushi or yakisoba instead of a Big Mac and fries? When Swedes come to Miami, are they open to attending a jai alai game rather than watching hockey as they are used to? Opportunities to try something new occur frequently in foreign cultures, and individuals who are adventurous enough to try new foods and activities are much more likely to adjust effectively. Families will not find the foods they are used to, but discovering new foods and family activities can be fun. The new culture will not be home, but if families can live with that, they will discover the new country's charm.

Stability

When individuals enter a new culture, a tremendous amount of stress may accompany the tidal wave of new experiences. Being able to cope effectively can be a significant buffer against these stressful experiences. For example, one study found that well-adjusted international assignees developed "stability zones" that functioned like harbors in a storm (Mendenhall & Oddou, 1985). These zones included such activities as engaging in hobbies, writing in diaries, and participating in contemplative or religious worship. The activities allowed the managers to withdraw temporarily from stressful situations and gain a better perspective on the new culture. The managers were able to break away from the constant struggle of trying to solve complex business problems made even more formidable because they did not fully understand the new culture's language, business customs, political systems, laws, and people.

Gender-Related Factors

So far, we have been discussing the importance of technical, strategic, communication, and individual factors in selecting individuals for global assignments. Some U.S. firms also pay significant attention to whether the candidate is a man or a woman. Fewer than 15% of U.S. human resource directors publicly acknowledge that they intentionally select male candidates more often than female candidates for global assignments, but the reality is that they do more than 90% of the time (Black & Gregersen, 1991a; Lineham, 2000; Stroh et al., 2000).

Recently, the International Personnel Association (IPA), whose members represent 60 of the top 100 multinationals in the United States and Canada, undertook a research study to find out why so few women are going on global assignments (Stroh et al., 2000). Stroh et al. interviewed the membership of the association and sent surveys to female international assignees and repatriates as well as to home-based employees at each company represented in the association. The study revealed that many human resource managers believed that women were less willing to accept international assignments than their male colleagues, would have greater difficulty than their male counterparts on global assignments, and would not be as well accepted as men in other cultures. Those women who had been or were on global assignments noted that in fact, they had accepted their assignments willingly, thought they had no more difficulty working in other cultures than did male managers, and felt they were as accepted in the countries where they did or were doing their assignments as men would have been.

Company-based data support these women's claims. When offered an international assignment, women accept about 90% of the time—that is, at the same rate as men. As a result of this important finding, companies that routinely eliminated women from consideration for global assignments are now including them in the candidate pool (Stroh et al., 2000).

Another survey by the Employee Relocation Council (1997) had similar results. In the 162 member organizations that responded, on average about 10% of their total international assignee population was female. Nearly 90% of the organizations indicated that females comprised 25% or less of their international transferees, and nearly one fourth had no female international assignees at all.

This bias also exists in Japanese and Finnish selection practices (99% and 91%, respectively).

Our research in the United States, Japan, and Finland as well as research by others indicates that women perform as well as men do both during and after global assignments (Adler, 1987; Adler & Izraeli, 1988; Black & Gregersen, 1991b). This result is even true for female international assignees in traditionally male-dominated societies such as Japan and Korea. For example, the Japanese view a female executive from Microsoft's corporate headquarters in the United States first as a company representative, second as a foreigner, and third as a woman. The first two factors render the international assignee's gender a nonissue for most Japanese businessmen.

Even if employees are not used to having women in the workplace, this cultural bias does not necessarily result in performance problems for women managers from other countries. Research has shown that in general, women managers exhibit many exceptional qualities for managing internationally (Stroh & Caligiuri, 1998).

U.S. firms should not discount the positive impact female international assignees can have on profits. If firms hope to select the very best candidates, they should cast an increasingly wider selection net throughout the company. This requires that more serious consideration be given to women. The increased level of global competition demands that firms discard unfounded biases and assess potential female candidates seriously when making international selection decisions.

EVALUATING FACTORS AFFECTING SELECTION

After a firm decides which selection criteria are most relevant to a global assignment, the next step is to determine how to evaluate candidates effectively on those criteria. Managers have a variety of tools to do such an assessment. Each of these tools has strengths and weaknesses, which are summarized in the answer to the following question: Is the tool reliable and valid?

A selection tool is reliable if it produces similar results in the hands of different people or at different times. For example, if both the human resource department and the line manager have interviewed a candidate for a global assignment and agreed that the candidate has strong communication skills, the method is deemed reliable. If a candidate completes a cross-cultural skills test on two different days and receives very different scores each time, the test is unreliable.

The validity of a selection tool depends on the extent to which the tool consistently finds that a particular selection factor is predictive of success during a global assignment. For example, if language skills are deemed relevant to a particular global assignment, then candidates may be assessed with a standardized language test. Even if this test produces reliable or consistent results for a candidate's language ability, it will be considered valid only if variations in test scores predict variations in success on the global assignment.

METHODS OF SELECTION

U.S. firms tend to rely on a very limited range of selection tools, but many tools are available. Some of the most effective tools are biographical data, standardized tests, work samples, and assessments at specialized centers. Selection interviews and personal references are widely used but are less effective. We discuss each of these tools and examine how reliable and valid they are for selecting candidates for international assignments.

Biographical and Background Data

The purpose of this selection tool is to gather background information about a candidates' personal and work histories. For example, professional or technical skills are an important selection criterion. These skills can be assessed reliably by reviewing a candidate's history. In the Mathison case, Doug Sweedlow's selection of Mike Tompkins was based in part on background data, which indicated that Mike had significant experience in information technology, an area directly related to the problems the Brazilian group was encountering. Whereas lack of technical experience can contribute to failure in a global assignment, biographical data (e.g., age, gender, race, work experience) and background data (e.g., past jobs and positions) are not strong predictors of success.

Standardized Tests

Standardized tests can be both reliable and valid methods of screening candidates for international assignments. For example, engineers are often required to take standardized tests for certification in different states or countries throughout their careers. These tests are usually quite reliable and valid predictors of an engineer's knowledge base.

As mentioned previously, research has demonstrated that certain personality traits are related to success in an international assignment. These traits, such as relational skills and extracultural openness, have been shown to predict performance in international assignments (Caligiuri, 2000). Unfortunately, U.S. firms, unlike their foreign counterparts, rarely use standardized tests to assess such criteria.

In our own research and work with multinational firms, we have developed a standardized test that assesses several important selection criteria for global assignments. Called the Global Assignment Preparedness Survey (G–A–P–S™), this selection mechanism appraises candidates for six criteria: cultural flexibility, willingness to communicate, ability to develop social relationships, perceptual abilities, conflict-resolution style, and leadership style (Black, Gregersen, Mendenhall, & Stroh, 1999). Our research has found that G–A–P–S results are related to a variety of outcomes including expatriates' work- and nonwork-related adjustment, job performance, satisfaction, and level of commitment and loyalty.

Many companies using G–A–P–S have found the individual feedback report that is generated to be extremely useful for "self-selection." Rather than using the G–A–P–S results to select and eliminate individuals under consideration for a global assignment, companies send the feedback to employees who have taken the survey to help them in deciding whether to even pursue an international assignment. The report contains both quantitative feedback that gives individuals a profile of their strengths and weaknesses relative to an international assignment and an idea of the strength and weakness of these attributes overall. For individuals who were choosing to go overseas for the wrong reasons (such as for financial reasons or because they had no more career prospects at home) and who were ill suited for a assignment, the G–A–P–S feedback report often helps them realize that going overseas may not be such a good idea. For others who have characteristics that make them more likely to succeed, the comprehensive nature of the survey and feedback reports help them focus on areas that need improvement. Some companies have also had partners complete a modified version of the G–A–P–S survey to help them self-assess their cross-cultural strengths and weaknesses. Although we do not recommend that G–A–P–S be used as the sole selection mechanism, it can be a reliable and valid component of an overall selection process. Some firms, especially in Asia, ask large numbers of young managers to complete the survey as a way to identify potential candidates for international assignments. Employees with reasonable prospects then have years, not just a few weeks or months, to improve their cross-cultural skills and leverage their strengths.

Work Samples

The goal of this selection tool is to place a candidate in a cross-cultural work situation as a way of evaluating his or her preparedness for taking on specific cross-cultural tasks. For example, one function of the position Mathison was establishing in Brazil was to chair meetings. To determine whether Mike was the best candidate for the position, it might have been useful to simulate a Brazilian business meeting by having some of Mathison's Brazilian IT employees and managers travel to the United States or by having Mike travel to Brazil and direct a meeting there. Although expensive, this method can be both reliable and valid. For example, Mike might well have had difficulty chairing such a meeting because he was used to working with employees in California. This approach probably would not have worked as well with Brazilian professionals. For example, there is great respect throughout Brazilian culture for rules and regulations, and people are generally resistant to change. Suggesting new ways to approach problems might not have gone over as well. Mike's performance in this work situation would have provided Doug with reliable and valid information about his ability to manage in Brazil.

Interviews

Of all the selection tools U.S., Japanese, Western European, and Scandinavian multinationals use in evaluating candidates for international assignments, interviews are the most consistently used (Björkman & Gertsen, 1992; Katz & Seifer, 1996). Unfortunately, an unstructured interview is not a highly reliable or valid method for effectively evaluating selection criteria (Hall & Goodale, 1986). For an interview to be valid, it needs to be structured in advance and behaviorally focused. In other words, the dimensions to be assessed must be predetermined and defined. In addition, the interview should focus on past behaviors that provide evidence of the presence or absence of the characteristics the interviewer is trying to evaluate. For example, rather than simply asking the candidate "Are you flexible?," the interviewer would ask about past behaviors. For example, the interviewer might ask "Describe what you did the last time you had plans in place for a project, things were going fine, and then a key factor outside of your control went in a direction you didn't expect." Firms such as Cendant Intercultural have well-defined behavioral interview techniques, which firms such as General Motors have used with great success in selecting expatriate managers.

WHO SHOULD EVALUATE CANDIDATES?

Decision makers must pay attention not only to selection criteria and methods but also to who performs the evaluations. In most U.S. firms, a limited set of decision makers selects candidates for global assignments. Most often, only one individual, the line manager responsible for the international unit, makes the decision, although others may also be involved in the selection process (e.g., representatives from the international unit if the assignment is to a middle-level or upper level managerial position). The human resource department is often underutilized; it usually plays an after-the-fact role making logistic and compensation arrangements. If firms want to be more strategic in their selection processes, they must learn to incorporate human resource departments' knowledge with that of line managers in home and host countries. In fact, human resource departments could act as decision hubs to ensure that a range of selection criteria are proposed, a variety of selection methods are utilized, and a full complement of candidates is considered.

WHO SHOULD BE EVALUATED?

Deciding whom to evaluate for a global assignment is not as easy as you might think. Because about 70% of international assignees around the world are married, most selection decisions involve not only potential international assignees but also the assignee's spouse. On one hand, inquiring into family matters is a delicate situation. On the other hand, as we discuss later in great detail, a family's willingness to relocate internationally and the employee's ability to complete the assignment successfully can have a significant impact on whether the employee performs effectively and completes the assignment. For this reason, firms must obtain some level of accurate information about the family situation. A firm should be straightforward in communicating the pros and cons of a global assignment through an interview or briefing, not over cocktails. A more formal setting provides an opportunity to make a more informed decision. Our research has found that when firms actively and directly seek partners' opinions about global assignments, partners are much more likely to adjust to interacting with host-country nationals and to living in their new culture.

WHY DO CANDIDATES ACCEPT OR REFUSE ASSIGNMENTS?

We have been focusing on the firm's approach to the selection process; however, the candidate's perspective is equally important. What factors influence a candidate's decision to accept or refuse a global assignment?

When considering an offer of a global assignment, candidates ask themselves two fundamental career questions: Will the assignment put me in a strategic business role, and will the assignment lead to my advancement? Because global managers, on average, have more than 14 years of experience in their parent companies, they want to be certain that an overseas assignment does not leave them out of sight and out of mind. A good indicator that a global manager will not be forgotten is the firm's clear vision about the strategic importance of an assignment and about how its success will produce tangible results and lead to upper management visibility for the manager. If the firm takes a "put out the fire" approach to global assignments, it will be hard to convince candidates that such assignments present long-term career advantages.

A global assignment can be expected to lead to advancement in the organization if the assignment is clearly defined as strategic before the selection decision is made. In other words, if an assignment is strategically important to the firm's success, then the international assignee has a much higher probability of being promoted after returning home. Our research shows that relatively few global assignments have led to promotions after the return home, even though most managers returning from abroad expect to get a promotion (Black & Gregersen, 1991b; Stroh, 1995; Stroh, Dennis, & Cramer, 1994). Only 11% of Americans, 10% of Japanese, and 25% of Finns received promotions after completing n global assignments lasting at least 2 years. Additionally, 77% of Americans, 43% of Japanese, and 54% of Finns were given positions lower than the ones held overseas.

Other studies corroborate these results, although some executives attempt to convince the business world that global experience does matter in today's multinational firms. Studies of Fortune 500 firms have found that executives only rarely pay attention to global experience as an important criterion in making promotions, and research by the Conference Board indicates that 80% of repatriates believe their international experience is not valued by their current companies. That same study reported that only 49% of participating companies even discussed the promotion with international assignees before departure. Furthermore, 87% of the responding companies reported that a majority of repatriates do not receive promotions on return. With these dismal findings, it isn't surprising that 25% to 50% of repatriates leave their companies on returning from a stint abroad (Gates, 1996; Stroh, 1995; Stroh & Lautzenhiser, 1994).

Collectively, these statistics paint a grim picture for potential international assignees and are very important in the selection process because candidates for the next set of global assignments are likely to see that returning expatriates are not promoted and may even be demoted. If this is indeed the case, it is quite unlikely that the selection process will result in the best candidates being posted; the best candidates will not want to jeopardize their careers with moves that result in demotion more than half the time. Firms must pay better attention to how they communicate through words and actions that global assignments really do count.

Financial incentives have played a significant role in attracting individuals to global assignments; candidates are interested in how a global assignment will affect their overall living situation (Stroh, 1995; Stroh et al., 1994). In Finland, for example, the high cost of new cars leads many individuals to leave the country on overseas assignments to avoid high import taxes and bring home a new automobile. In the past, a global assignment was often seen as an opportunity to live like royalty for several years. This corporate emphasis on the financial rewards of global assignments has created a strong expectation on the part of upcoming expatriates that they have the right to luxury treatment; however, the increasingly competitive global business environment has forced most firms throughout the world to reduce these financial incentives. This trend reinforces the need for firms to clarify the nonfinancial benefits of global assignments such as completing a strategically important mission for the firm.

Finally, many candidates consider the learning aspect a critical component in their decision to accept a global assignment. When asked why they would accept such an assignment, more than half of future global leaders cited the opportunity for personal growth and to gain cross-cultural experience.

FAMILY'S ROLE IN SELECTION AND SUCCESS

A disappointing scene is being played out in more and more multinational organizations just as they have begun to recognize the importance of expanding their global operations. An executive offers a plum overseas assignment to the best candidate in the company only to be turned down because the candidate's spouse or partner is unable or unwilling to put a career on hold while accompanying an expatriate on a global assignment. What can the company do? On one hand, management recognizes the need to compete in the international arena. On the other hand, management is aware of the stress and disruption relocation can cause a family.

As often as this scene is occurring today, we can expect it to occur even more frequently in the future. Currently, 80% of all couples are in dual-career marriages. Clearly, family conflicts are already a concern for companies as they try to expand their global operations. A study by the IPA (Stroh & Caligiuri, 1998) indicated that although management in multinational corporations consistently rank developing global leaders as a high priority, because of family considerations, notably dual career marriages/partners, multinational companies are struggling to find better ways to manage their dual-career employees' international assignments.

Some corporations are beginning to involve partners and spouses more actively from the time a manager is being considered for an assignment abroad. The goal, of course, is to increase the chances that global assignments will be completed and that the work will be performed effectively. Corporations are beginning to recognize that there is a close link between achieving these goals and how well the family adjusts to their new location. In the next section, we discuss an approach to the selection process that emphasizes this link.

FAMILY SYSTEMS APPROACH TO SELECTION PROCESS

Research and experience have shown that the success of a global assignment is highly dependent on the attitudes of a manager's family at the time an offer is made to relocate and the ability of the family to adjust during the global assignment. To ensure the greatest chances of success in the highly competitive global arena, organizations need to evaluate potential candidates for global assignments on much more than just their managerial skills and experience. They need to recognize the role of the entire family unit in determining whether the manager will (a) accept the offer, (b) adapt successfully to life and work in the foreign location, and (c) complete the assignment. Evaluating the candidate from a systems perspective that includes the family as an integral part in the selection process increases the odds of achieving this goal.

From a systems perspective, we recognize that an individual's actions are influenced by the other members of his or her family as well as by the individual's own past actions (Stroh, 1995). Seen from this perspective, the potential international manager is viewed as a subsystem of the family, influenced particularly by a partner's willingness to relocate internationally and by the family's ability to adjust while abroad. Couples develop patterns of interaction that enable each partner to support the other. These patterns develop through processes of mutual accommodation and influence. An opportunity to relocate abroad is not going to dislodge this process. Consequently, it's unwise for those selecting international assignees to think that work and family life are separate or to disregard the role or influence of the spouse on the success of the assignment.

Factoring in the Partner's Opinion

Several studies have shown that a spouse's attitude toward relocating internationally is the most important factor in predicting a manager's willingness to accept an offer of a global assignment. Among the researchers who have studied this issue are Brett and Stroh (1995). Brett and Stroh surveyed 518 male and female managers and their spouses from 20 multinational U.S.-based corporations to examine this issue. Brett and Stroh found that although a manager's own attitude toward relocating internationally had a powerful effect on the decision to accept or reject an offer, the spouse's willingness was even more important. In other words, when the manager's spouse or partner is unwilling to relocate, the manager is likely to reject the offer. Partners often feel that putting their careers on hold will jeopardize their opportunities for future career advancement. In some studies, the primary reason managers gave for turning down offers was the negative effect the move would have on their partners' careers.

At the same time, executives should not eliminate anyone from the pool of potential candidates based on whether a manager's partner has a career of his or her own. Short-circuiting the decision to accept an international relocation by presuming that a manager in a dual-career relationship will turn down the assignment or not succeed in the assignment is not advisable for at least two reasons. First, firms need their best and brightest people to gain international experience and cannot afford to remove employees in significant relationships from consideration (remember, about 80% of all married individuals are part of a dual-career couple). Second, research has shown that demographic characteristics (whether the employee's partner has a career; employee's age, gender, or education) are not good predictors of willingness to relocate internationally (Brett & Stroh, 1995). In essence, as Brett and Stroh (1995) noted, "What seems obvious may not be true." The best way to know whether a candidate will accept an assignment is to ask the candidate.

Whether a dual-career couple will accept an offer to relocate internationally is not consistently an issue in multinationals across the globe. In Japan, for example, a spouse's willingness to relocate has less influence on the employee's decision to accept or reject an offer than in Western countries such as the United States. This is partly because fewer employees are part of dual-career couples in Japan and partly because decisions and roles relative to work and family are more separated. The breadwinner (usually the man) makes decisions about work, and the spouse (usually the woman) makes decisions about the house, the children's education, and other matters including investments. However, gender-specific behavior in Japanese families (men = work; women = home and children) is beginning to break down as the number of dual-career couples in the Japanese workforce increases.

Income-Related Concerns

Families dependent on two incomes may also be concerned about the effect an international transfer will have on their standard of living, especially because the employee's partner may not be able to obtain a work permit in the country to which they relocate. Thus, not only are partners often forced to put their careers on hold while they are abroad, but their loss of income may also have serious implications for the family's overall financial future.

Other Family-Related Considerations

Managers often anticipate difficulties in providing for their children's education and safety in international settings and may be reluctant to accept an offer to relocate internationally for these reasons. Managers whose children have special education or health problems may rule out international relocations completely.

According to one study, after a partner's career issues, meeting special family needs (e.g., educational, medical, or social) and concerns about caring for elderly relatives were the main reasons managers turned down offers to relocate or removed their names from consideration (Brett & Stroh, 1995). Corroborating these findings, another study (Windham International, National Foreign Trade Council, & Society for Human Resource Management, 2002) found that a candidate's unwillingness to accept an overseas assignment was related to family issues 86% of the time.

WAYS COMPANIES CAN ADDRESS FAMILY CONCERNS

Progressive companies are beginning to recognize that "sweetening the pot" for managers and their families can be an effective way to increase the likelihood that managers will accept offers to relocate internationally, thereby enabling the company to better compete in the international arena. In particular, a growing number of companies are compensating spouses for lost income. In one study (Thompson, 1998), for example, 163 companies in 20 industries compensated spouses in this way. Another potential resolution to a family's unwillingness to accept an international assignment is to consider a commuter relationship in which the employee commutes between their home and the international assignment. Although this may seem like a tremendous burden, with many company and family details to work out, many international assignees consider using this commuting relationship as an acceptable compromise (Stroh, 1999). In the following section, we discuss relocation assistance programs six companies have established.[1]

Case 6: Sara Lee

The Sara Lee Corporation has what it calls "Work and Lifestyle Assistance" benefits. Unlike traditional spousal assistance programs, Sara Lee's program offers benefits not only to spouses but to single parents to be used for child care, for families that care for disabled or elderly relatives, and to employees in a variety of other circumstances. The important aspect here is that Sara Lee very broadly defines family to include a broad form (even including singles). The hope is that by offering these benefits, expatriates will be able to adjust more easily to their new positions and cultures.

At Sara Lee, the human resource department has established links with local employers in countries where Sara Lee has international offices. These contacts are valuable in finding employment for the spouses of Sara Lee employees. Staff also investigate employment opportunities within Sara Lee. If the company is unable to find the spouse employment, the spouse is given up to $5,000 per relocation to help her or him conduct a job search.

Case 7: Quaker Oats

Quaker Oats has similar policies and reimburses spouses up to $5,000 during the assignment for expenses related to seeking employment; however, its policy does not state that the spouse must be unemployed. Eligible expenses include but are not limited to (a) expenses incurred in obtaining a visa or work permit, (b) assistance in producing a resume and submitting it to appropriate employment agencies, (c) career guidance and consultation, and (d) continuing education activities. Both Quaker Oats and Sara Lee also have policies aimed at meeting the needs of single employees, single-parent families, and expatriates with elderly or disabled dependents.

Case 8: Eastman Chemical

Eastman Chemical Company established its initial policies on spousal benefits when the company was created in 1994 as a spin-off of Kodak. Although it has

---

[1]Much of the data for the case studies in the next section were obtained by Stacey Bogumil, Kelly Chew, Antoinette Cromartie, and Brooke Zahara as part of a group research project they did for a human resources class (HRIR 443) at Loyola University Chicago. We appreciate their contribution.

made some changes in these benefits, it has not felt the need to sweeten the pot further.

Eastman Chemical offers spouses three separate payments based on the spouse's annual income before relocating. The 1st-year payment is 33% of the spouse's annual income before the move; the 2nd-year payment is 67% of the income; and the 3rd-year payment is 33%. There is a cap of $10,000 on the payments, and spouses are eligible for this assistance only if the spouse was forced to give up her or his job to relocate. Spouses who find a job are also eligible for assistance so long as the spouse's salary is less than he or she was making at home.

Eastman Chemical must approve how a spouse chooses to use the annual payments. Spouses typically use the extra money to obtain assistance in preparing a resume or job search, to attend classes, or to recertify professional licenses.

Spouses of Eastman Chemical expatriates universally take advantage of the benefits the company offers. So far, not a single spouse has declined the assistance.

Progressive policies like Eastman Chemical's are still uncommon. Eighty-eight percent of multinational companies provide no such adjustment, 2% do, and 10% make other arrangements negotiated case by case.

Case 9: Colgate-Palmolive

Colgate-Palmolive has an attractive reimbursement program that pays spouses up to $7,500 over the course of an international assignment. The program is quite liberal in what it will reimburse so that someone who is changing careers may well be reimbursed for courses related to their new profession. Spouses may also receive up to $10,000 per year from the company's tuition assistance program.

Case 10: Coca-Cola Enterprises

Coca-Cola Enterprises has a spousal assistance program to help spouses further their education or professional development. The program, which provides $2,500 per year for 3 years, is available to spouses who gave up jobs to relocate with their wife or husband. Spouses accompanying their husbands or wives on 1-year assignments are entitled to $1,000.

Case 11: Monsanto

Since January 1, 2002, Monsanto has had a clearer, more broadly defined spousal benefit policy than its earlier one. In particular, Monsanto's current policy covers both spouses and domestic partners. This was in keeping not only with best practices but also with Monsanto's benefits plans. The company also eliminated employment status as a criterion for coverage, thereby entitling stay-at-home and unemployed partners to receive benefits. Partners are entitled to up to $5,000 per year for the duration of the international assignment to reimburse them for expenses associated with adjusting to the new culture or finding a job. Among the more popular ways partners use the program is for reimbursement for the purchase of a personal computer.

Case 12: Motorola

Motorola also revamped its spousal benefit policy effective January 1, 2002. The company offers spouses of international assignees $5,000 per year for up to 6 years, allowing for the possibility of back-to-back international assignments. Like Monsanto's policy, Motorola's covers both career- and noncareer-related expenses including membership fees to join clubs and golf and language lessons.

The Motorola Company has been a vanguard in developing spousal assistance policies. The company's philosophy is that if a company expects its business to be on the cutting edge of the industry, its human resource policies must also be on the cutting edge. Not only was Motorola among the first organizations to offer spousal assistance policies, but it was among the first organizations to evaluate their effectiveness.

Surprisingly, until recently, few Motorola employees were using the program. A study of its spousal assistance programs in China, Hong Kong, Singapore, Japan, the United Kingdom, and the United States revealed that poor communication about the policy and its availability contributed to underuse. Today, employees *and their* partners *are notified of the plan during preassignment counseling. They* may *also access policy guidelines via the* company's *Web site. Based on company surveys, Motorola has determined that both international assignees and their partners have high regard for the spousal assistance program. Most noted that the program has helped with both cultural assimilation and repatriation.*

Spousal assistance programs like Motorola's are an excellent way for companies to help managers' spouses make the difficult adjustment to a foreign location. Because a spouse's adjustment is positively related to the employee's adjustment, instituting such programs should lead to higher productivity and greater likelihood that global assignments will be completed. Of the Motorola spouses who took advantage of the program, 61% said it helped make their adjustment easier. Further, 75% claimed the program also contributed to their partners' successful adjustment (Pellico & Stroh, 1997).

Participation in a spousal assistance program also pays off for spouses on repatriation. Of the spouses who took advantage of Motorola's spousal assistance program, 79% percent said that they felt better prepared to enter the workforce when they returned home. As one spouse commented, "I have been out of the *workforce for four years; I think it will look better to an employer to see I was doing something constructive" (Pellico & Stroh, 1997).*

APPROACH TO GETTING THE RIGHT PEOPLE

This section contains recommendations for how multinational firms can strategically approach the selection process for global assignments. The first stage of this process is summarized in Fig. 3.1.

Strategic Analysis of Global Assignments

Making global assignments truly strategic requires the foresight to perform a careful analysis of the firm's overall global assignment needs, define its current global candidate pool, and most important, assess whether its pool of candidates will be large enough to meet future demands for effective global managers.

Analyze Current Needs. The analysis of current needs should consider several critical factors. What stage of globalization is the firm in? Companies in the export stage have significantly fewer demands for international employees than do firms in the coordinated multinational stage. Another important consideration in assessing the global assignment needs is the strategic functions that such assignments should play. Does the firm need to send people from headquarters for purposes of coordination and control? Does the company need to raise the level of communication between headquarters or between subsidiaries? Does the firm need to develop more future executives by giving them global assignments as developmental experiences? Answers to these questions can help a firm decide, from a strategic perspective, what the current needs actually are.

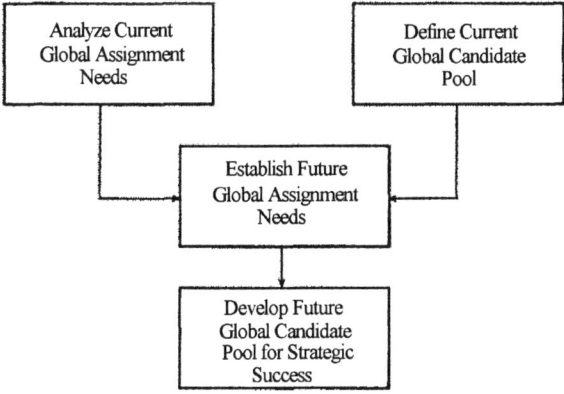

FIG. 3.1. Strategic analysis of global assignments.

Define Candidates. Multinational companies need to know the composition of their current candidate pool of global managers. For many companies, this pool is a "black box" because there is no centralized clearinghouse for collecting and updating information on candidates with relevant skills and linking these individuals with potential needs. Some companies such as Neste Oy, a major Finnish oil and gas firm, have developed comprehensive databases that detail a manager's current assignment, technical qualifications, previous global experience, cross-cultural skills, and management potential within a firm. Such data require an initial investment of time and resources; but they can be invaluable in searching for the candidates with the best technical and cross-cultural skills to staff a specific global position.

Assess Future Needs. Firms must plan for the future by deciding what their future global assignment needs will be. Again, these needs will be a function of a firm's future stage of globalization and the necessary strategic functions to sustain an advantage. If a firm is currently at the multidomestic stage with uncoordinated operations in two or three countries but intends to become a coordinated multinational with operations in several more countries, that company will have an increasing number of global positions to fill.

In addition to its stage of globalization, a firm must consider what its key strategic functions will be in the future to assess future needs accurately. For example, if a firm intends to make acquisitions throughout the world to develop additional technological synergies, it will need to move technology and information from operation to operation and from overseas to headquarters. The effective flow of information may well require more global assignments.

Develop the Candidate Pool. The final strategic step in preparing for the future is the development of a firm's candidate pool. To develop a sufficient pool of qualified candidates for global assignments, a company must implement regular assessments of employees' managerial and cross-cultural skills. In addition to examining managerial advancement potential through traditional succession-planning mechanisms, the firm should regularly assess a variety of skills and individual characteristics associated with successful global assignments including communication skills, conflict-resolution skills, leadership style, foreign-language skills, stress-reduction capacity, and cultural flexibility. An analysis of these important cross-cultural skills could be incorporated into traditional assessment center programs or management-training courses. The skills could also be assessed with surveys such as G–A–P–S.

In addition to regularly assessing managerial and cross-cultural skills, firms should create strategies and plans for systematically developing skills in which many employees may be weak. For example, the Lord Corporation, a medium-sized, privately held U.S. manufacturing firm, was preparing to set up production operations in France, so the company offered free French-language classes on company time to interested employees. Lord also held "French Day" at corporate headquarters once a week. On French Day, the corporate cafeteria served French foods so that people could try previously unfamiliar gastronomic delights. This relatively simple but strategically thought-out tradition helped many employees develop greater cultural flexibility and improve their spoken French. These activities may seem minor, but they help communicate the genuine importance of global competence to employees and provide them with opportunities to develop the skills necessary to complete international assignments.

Selecting Candidates for Specific Assignments

After conducting a strategic analysis of global assignments within a firm, managers must face the reality of selecting appropriate individuals for specific positions. To assist managers in the decision-making process, Fig. 3.2 demonstrates a flow chart of key activities that should lead to more successful and strategic outcomes.

Form a Selection Team. The first step in the selection process is to create a selection team. This team should include at least three members: a home-country manager, a host-country manager, and a human resource department representative. The home- and host-country managers help ensure that headquarters and

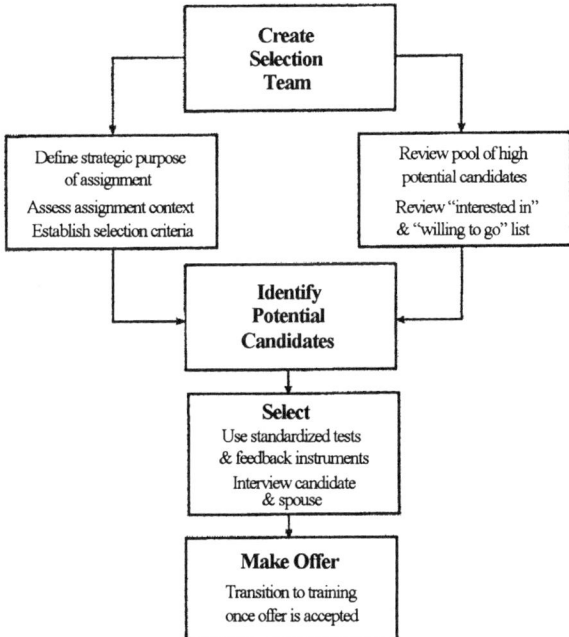

FIG. 3.2. Global assignment selection process.

subsidiaries are both served in the selection process. Furthermore, the home-country manager might be designated as the expatriate's "sponsor" for the global assignment. The human resource representative can serve several important functions for the selection team such as ensuring that a range of selection criteria are utilized and helping to locate a broad slate of candidates for the position.

Define Strategic Purpose of Assignment. The next step is for the team to determine the strategic purpose of the global assignment. Currently, the goal of most assignments is short-term problem solving. Firms need to be more reflective in deciding the strategic functions of the assignment before the assignment is made.

Assess the Context. What is the cultural context of the country where the company is sending the employee? If an assignment requires extensive interaction with host-country nationals, cross-cultural communication and language skills will be important. If the general culture of the host country will be unique and therefore more difficult to adjust to, this factor may have a significant impact on the selection criteria and decision.

Establish Selection Criteria. The selection team needs to define the technical criteria of the job as well as the strategic functions of the assignment and the cross-cultural context of the position. For example, specific engineering knowledge may be needed; if the assignment is also developmental, then choosing someone with the potential for advancement as a manager is crucial. If extensive interaction with host-country nationals is also required, the selection team should pay attention to cross-cultural communication skills. Finally, the more difficult the general culture of the foreign country, the more attention the selection team should pay to issues such as flexibility and ethnocentricity.

Review the Pool of Candidates. To ensure that those being considered are not simply the employees that a given individual happens to know, many leading-edge companies have pools of high-potential candidates. The companies survey employees every couple of years to identify who is interested in and willing to go on international assignments. In global companies such as Colgate-Palmolive, the lists of high potentials and those interested in international assignments are nearly identical. However, because of individual circumstances, not everyone in the high-potential and interested group is willing to go at the moment. This information allows Colgate to avoid interviewing candidates or making offers to candidates who cannot accept now. This approach also helps avoid the negative effects on the candidate of appearing to turn down an international assignment.

Define the Candidate Pool. Once appropriate selection criteria have been developed, the firm can utilize references, internal job postings, and a global candidate pool database (if available) to match the highest number of potential candidates with a particular assignment.

Administer Tests and Feedback Instruments. After a candidate pool has been defined, the human resource member of the selection team can facilitate the use of a variety of selection methods. An overseas assignment will be both costly and risky, so the expense of utilizing a standardized feedback instrument such as G–A–P–S to ensure a sound selection process is an investment in the future rather than just an immediate expense. Because most decision makers tend to use selection methods that are unreliable or invalid (such as one-on-one interviews), the selection team should decide which methods are most effective for evaluating which selection criteria for a particular assignment.

Interview Candidates and Spouses. At this stage, the selection team will have narrowed the field to one or two potential candidates who, it is hoped, have both the technical and the cross-cultural qualifications to succeed in the global assignment. An in-depth interview that outlines the strategic purpose of the assignment and its relationship to the candidate's career path within the firm and that includes an honest assessment of life in the foreign country can provide a realistic preview for the candidate. In addition, interviews or briefing sessions with the spouse to provide him or her with realistic expectations about life in the foreign country and to determine unique dual-career and family needs can significantly enhance the chances for success. The importance of providing a realistic preview of the job and of living conditions cannot be overemphasized. Interviews must be conducted in a context in which both the organization representative and the candidate can honestly share perspectives on overall aspects of the job and the foreign country. In this chapter's "From the Front Lines," an executive at Archer Daniels Midland (ADM) describes some of the ways that company addresses adjustment issues before an employee has even accepted an offer to go abroad.

Make the Offer. If the selection team approves the candidate after utilizing several selection methods and conducting interviews with the candidate and the spouse, and if the candidate and the spouse are favorable to the assignment after having been given a realistic preview, an offer can be made. This decision will be based on more relevant, factual, and comprehensive information than most companies currently use to select individuals for global assignments.

Make the Transition to Training and Preparation. The final stage of the selection process entails the transition from acceptance of the assignment to preparation for it. Generally, if the selection process has been strategic and not tactical, the assignment will be made far enough in advance for appropriate training and preparation to be initiated.

FROM THE FRONT LINES:

"ADM's Strategies for Addressing Spouses' Adjustment Problems"
by Maureen Ausura, Corporate Vice President,
Human Resources, Archer Daniels Midland

At ADM, we have found that it is usually the spouse of the ADM employee who initiates the decision to return home from a global assignment. The employee—almost always a man—is excited about the opportunity to spend time abroad and sees it as a way to advance his career. When he gets to the new location, he immediately becomes involved in his work. The children are in school and quickly make friends. However, the spouse is typically at home—in a new community in an unfamiliar culture. Frequently, she does not speak the local language, is usually unable to work, and in many of our locations, cannot safely go out of the house because of concerns about security.

ADM has begun several initiatives designed to minimize the adjustment problems many spouses experience. Among the strengths of our program is that we discuss adjustment issues with spouses even before employees have decided to relocate abroad.

Predeparture, the family meets with a counselor from our employee assistance program who discusses adjustment issues and assesses the spouse's willingness and ability to adapt to the culture in the country to which they are planning to move. Since developing this initiative, one employee turned down a global assignment after initially accepting it because of concern about his wife's ability to adjust.

ADM also offers customized destination services. These can range from help in finding housing and setting up bank accounts to advice on matters such as how to network in the community and become involved in special-interest groups.

In addition, ADM provides a $5,000 spousal-assistance supplement. This was designed to help the spouse find a job or set up a business. However, many spouses use the money to attend classes in the new community; some even pursue a degree. Finally, in "hot" locations, our security personnel hold seminars for both the international assignee and his spouse on how to protect themselves at home and while traveling.

We have found that the key to helping spouses adjust to living in a foreign country is to ensure that they become involved in their new communities and develop support networks. Our goal is to give them tools to assist them and therefore increase their odds of adjusting well to their new homes.

SELECTION DECISIONS: THE KEY TO FUTURE SUCCESS

Two fundamental points should be made about the selection process for global assignments. First, the selection process must be ahead of the firm's globalization process. In other words, as the firm advances from one stage to another stage of globalization (e.g., from a primarily export to a global firm), the firm's human resources function must already have a sufficient pool of international human resources or potentially successful international managers to sustain the next stage of global expansion. Second, acting strategically is the key to keeping ahead of the globalization process. Conducting strategic assessments of needs and developing a strong candidate pool can enhance the global competitiveness of the firm and the success of the individual. Otherwise, decision makers easily lapse into ineffective selection practices such as using technical qualifications as the only selection criterion or relying on interviews as the only selection method. By maintaining a strategic orientation at each stage of the selection process, the firm is much more likely to receive a positive return on its high-cost investment as it selects expatriates with the necessary technical and cross-cultural skills who can solve short-term problems and accomplish long-term strategic objectives.

CHAPTER

4

Training: Helping People Learn to Do the Right Things

Among the most challenging aspects of international human resource management is how to prepare international assignees to succeed abroad and once they are abroad, how to facilitate their development so that they may gain competencies as global managers. This chapter provides a framework and concrete guidelines for designing and implementing an effective training and development program for international assignees. These ideas are considered in the context of the challenge facing a fictional human resources (HR) vice president, Mel Stephens, who has to develop a training program for several employees being sent to Japan.

"I must be out of my league on this one," Mel thought. He had faced many challenges as vice president of human resources since arriving at Recor Engineering, but none had nagged at him so persistently. Recor Engineering, a leader in the U.S. domestic construction industry, had just sealed a joint-venture pact with one of Japan's largest construction firms, Dentsu Hogen K.K. Recol, a San Francisco-based company, and had agreed to send a large team of American experts to Osaka with a special group of Dentsu Hogen's best engineers. The Americans were to team up with their Japanese counterparts to bid on a project to expand the runway at Osaka Airport as well as related ventures.

All of Recor's engineers—a total of 18—had agreed to relocate to Japan after being assured that their families' financial positions and standards of living would not suffer as a result of the 3-year assignment. None of the engineers, most of whom were married, had indicated any reluctance on their spouses' part concerning the relocation. Nevertheless, Mel's secretary—his hidden ears in the company—had told him 3 weeks earlier that she knew of at least eight spouses who were "less than thrilled" about disrupting their children's education and relocating. Five of the spouses had also indicated that they were not pleased about having to quit their jobs to move overseas, even though the international assignment compensation package was lucrative.

Mel quickly discovered that guaranteeing his people good pay was much easier than his next task—deciding how much and what kind of predeparture training to give them. His phone calls to colleagues had yielded mixed responses. Some felt that no training was necessary, some felt a short "area briefing" was sufficient, and a few had heard of consulting firms that offered comprehensive training packages.

He followed up with a few of the consulting firms that his colleagues suggested, but the prices they quoted would push the upper limits of the quarterly training budget. Mel then talked with people at a variety of firms that had managers in Japan. None had provided any in-depth predeparture training.

Mel ejected his Mozart disc and found a news station on his car radio. He tried to relax as he listened to the news, but the nagging feeling of the past few months returned. "We're sending these guys into a strange, totally unfamiliar culture," he thought.

Or at least it seems strange to me. Shouldn't we do something to prepare them? Other firms don't do much, if anything . . . but if this joint venture melts down, I'll be in a tight spot. On the other hand, the people being sent have all been successful here. They should do fine. Besides, if they're worth what we're going to pay them, they should be able to work through whatever problems come up.

Mel put his thoughts on hold and turned his attention to the weather report.

Was Mel justified in being concerned that Recor's managers needed cross-cultural predeparture training? How much training do managers sent on global assignments usually receive? Management research has uncovered the following facts across industries:

1. Studies show that managers who are being relocated are often offered no cross-cultural training before their departure (Windham International et al., 2002). Studies differ on the figures, but several studies reveal that one third to one half (and in some cases as many as 90%) of international assignees receive no predeparture cross-cultural training (Marx, 1996; Mendenhall, Kuhlmann, & Stahl, 2001; Mendenhall, Kuhlmann, Stahl, & Osland, 2002; Suutari & Brewster, 2001; Suutari & Burch, 2001; Windham International, 2002).

2. Of the firms that do offer cross-cultural training, the most common training activities involve watching films, reading books, and talking with people who have lived in the country to which the manager is being sent. Few firms offer in-depth, rigorous, skill-centered, cross-cultural training (Brewster, 1991; Mendenhall et al., 2002; Oddou & Mendenhall, 1991; Tung, 1981).

With so many individuals and families receiving no training or less training than necessary, it is no wonder so many global managers struggle in their overseas assignments. The question is not whether firms should allocate time and resources to training their global managers but rather how firms should construct such training to ensure that it meets their global managers' needs. We do not advocate the use of "canned" programs; instead, a firm's thoughtful responses to a variety of training issues should drive the nature of the training offered.

UNDERSTANDING HOW PEOPLE LEARN AND ADAPT TO NEW CULTURES

An understanding of how individuals learn and adjust to new business and social cultures is critical to developing effective cross-cultural training. Much research, which is covered in detail in chapters 2 and 5 (this volume), has been done on this subject. The process of adapting to a new culture involves several learning principles that need to be accounted for when designing cross-cultural training (Bandura, 1977; Black & Mendenhall, 1990b; Manz & Sims, 1981).

Three-Step Learning Process

In this case, principles are presented as a series of learning steps. Associated with each step is an example from the actual experiences of an expatriate manager, Earl Boyer, who was in charge of a large ranching operation in an island in Polynesia.

Step 1: Attention. Before managers can alter their behavior so that it conforms to the norms of the host culture, they must first see, attend to, and become aware of how the locals behave. Typically, people need to view behavior and think about it before they can decide whether they want to try it out.

For example, the Polynesian employees of the ranching operation regularly held parties on the weekends. An integral part of the get-togethers was the *hangi,* which basically consists of cooking food on white-hot rocks in a hole in the ground. Flax leaves are placed over the heated rocks, then meat is laid on the flax, then more flax leaves are placed on top of the meat. The process is then repeated with a variety of vegetables. Finally, dirt is put over the food until the hole is filled in. Later, when the food is cooked, the "oven" is dug up and the food is distributed to everyone.

The Polynesian managers, supervisors, and workers viewed these get-togethers as culturally very important. A lot of singing, dancing, storytelling, and renewing of family and community ties took place. Being together and the warm feelings associated with being part of a social unit were important; people who did not come to the parties or who were unwilling to loosen their inhibitions and sing, dance, and so on were viewed as cold, aloof, and untrustworthy. The American manager, Earl Boyer, sensed all this and realized that attending the get-togethers as well as sponsoring them would be important if he was to be effective.

Step 2: Retention. During the second step, managers typically think about what they have learned, seen, or heard regarding culturally appropriate behavior in their new country. The more they think about what they have seen, the more they develop a "cognitive map" regarding when to produce certain behavior, under what circumstances it is acceptable and unacceptable, whether it is all right for foreigners to produce the behavior, the penalties for not producing the behavior, and so on. Examples of the behavior in question get locked into memory and become a reference point for understanding and reproducing the behavior. During the early part of the retention process, important new behaviors and the norms surrounding them must be quite conscious. Eventually, as the new behaviors and their norms are understood more completely, that knowledge settles into memory and spurs reactions in social and business situations naturally and unconsciously.

This process is not unlike driving in a city that's new to you. Without a map, you get lost easily. To avoid this, you look at the map often and keep it close while driving. Slowly, over time, you learn where streets are, the shortest routes to work, and so on until you no longer need the physical map because you've made a mental map. Likewise, in a foreign culture, with effort the culture becomes less new after a while and behaviors become more predictable because we develop cognitive "cultural" maps that inform us what to say and do. In the New Zealand example, Earl Boyer continued to attend the parties hosted by his workers and their relatives and carefully observed the process of the *hangi,* participating selectively when the Polynesians called on him to do so. Mainly, he blended into the group and carefully observed everything that was going on, developing an understanding of the rules, their purpose, and the behavior expected of participants. He observed that the food was an important stimulant to the social norms that caused cohesiveness in the work group and among members of their extended families and friendship networks.

Step 3: Trying Out the New Behavior. Basically, once cross-cultural behavior has been observed and managers have gained a mental understanding of the rules associated with the behavior, they must decide to try out or not to try out the behavior. As managers experiment with new behavior, they check their performance against their cognitive maps until they become expert at the behavior. If trying it out causes embarrassment or negative reactions from host nationals, the managers may never try the behavior again and thus risk never adjusting to their new culture. The closer managers' cognitive maps accurately reflect the culture, the more likely they will reproduce new behaviors successfully.

Having finally decided that he was ready to host a *hangi,* Earl Boyer invited all the Polynesians in the community to his house. When they arrived, they saw smoke rising out of the ground. The festivities began, and all were having a good time.

When the *hangi* was ready to be unearthed, the Polynesians began to ask each other who had helped their American boss host the party. After a while, it became obvious that no one had. All eyes focused on Earl, who simply smiled and yelled, "*Haere Mai Kita Kai!*" ("Welcome, come and get the food!"). His ability to influence and gain compliance from his workers increased dramatically—literally overnight—for no other American had ever attempted to prepare a *hangi* before.

Training and the Three-Step Learning Process

The case just presented highlights three areas on which training for global managers must focus—namely, helping managers to

1. Become aware that behaviors vary across cultures and the importance of observing these cultural differences carefully.
2. Build cognitive cultural maps so they can understand why the local people value certain behaviors and how those behaviors may be appropriately reproduced.
3. Practice the behaviors they will need to reproduce to be effective in their overseas assignments.

Without training, some global managers will succeed at negotiating the three-step learning process by themselves, but many won't. Good training can be a tremendous help in learning about new cultures.

Designing Cross-Cultural Training

Research shows that an important factor in the success of a cross-cultural training program is its rigor—the degree of mental involvement and effort that the trainer and the trainee must expend for the trainee to learn the required concepts. The ability of a firm to determine the degree of rigor that is appropriate is the key to the design of valid cross-cultural training (Black & Mendenhall, 1989; Mendenhall, Dunbar, & Oddou, 1987; Mendenhall & Oddou, 1986; Tung, 1981, 1982).

Low-rigor training includes activities such as watching films, listening to lectures and area briefings, and reading books. By contrast, more rigorous training requires the trainee to learn skills passively but also to practice them. More rigorous training would include role modeling, videotaped sessions to demonstrate success at mastering skills, and language training. High-rigor approaches extend the degree of participation on the part of the trainee through the use of assessment centers, interactive language training, and sophisticated cross-cultural simulations. Figure 4.1 illustrates the close relation between training rigor and participant involvement.

Rigor is also associated with the length of time spent on training. A training program that required trainees to participate for 25 hr over 5 days would be less rigor-

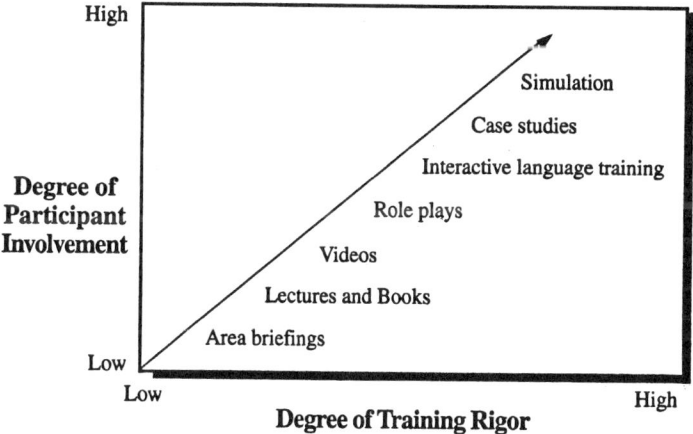

FIG. 4.1. Degree of training rigor.

ous than a program that spanned 100 total hours over 3 weeks. Even though the time and expense involved usually increases with the rigor of the training, so does the level of what is learned and retained. For example, people generally remember only about 17% of what they hear in a lecture. In contrast, retention from more rigorous training methods such as simulations is often in excess of 85%. Consequently, it is important not to cheat global managers by offering them "quick-and-dirty" programs. It's also important not to assume that just because a program is offered, it is effective. Viewing training as an expense rather than an investment often results in inadequate and inappropriate training. Training for international assignees and families is an investment that pays off best when the rigor of the training is appropriately matched to the nature and requirements of the assignment.

Once key decision makers understand the need to offer global managers rigorous training, the critical question is how rigorous the training should be. To help firms answer this question, we offer a framework that includes three dimensions that must be taken into account when deciding what methods should be included in the training design (Black, Mendenhall, & Oddou, 1991). These dimensions are the degrees of cultural toughness, communication toughness, and job toughness required of the manager in the overseas assignment. A careful analysis of these three dimensions is necessary before valid training can be designed. Otherwise, the wrong training will be given.

Cultural Toughness

Some cultures are more difficult or tough to adjust to than others. The underlying values that determine the way business is done are more different from American norms in some countries than in others. A manager going to Japan will have a tougher time, and the learning curve will be longer than a manager going to Australia. Both expatriates will encounter cross-cultural problems, but the challenges facing the manager in Japan will be more severe. The tougher the new culture is to adjust to, the more assistance, through rigorous training, the global manager will need to adjust and be effective. How does one determine which cultures are toughest? For example, research by Torbiörn (1982) suggests that the following regions of the world, in descending order of toughness, are the most difficult for Swedes to adjust to:

1. Africa.
2. Middle East.
3. Far East.
4. South America.
5. Eastern Europe/Russia.
6. Western Europe/Scandinavia.
7. Australia and New Zealand.

Individual countries within regions vary in cultural toughness, and there are within-country differences as well, but this simplification of the research provides a fairly good idea of the need, amount, and rigor of training necessary by region of assignment. "From the Front Lines" provides a real-life example of the value of cultural training.

The next factor that determines cultural toughness is the previous overseas experience of the specific candidate for the global assignment. The more experience the person has had with the specific culture to which he or she will be assigned, even if that experience was in the distant past, the more the person should be expected to be able to cope with the challenges in the new culture. Someone who has already lived and worked in Nairobi for 3 years and is reassigned there after being home for 5 years will need less rigorous training than someone who has never lived or worked in Nairobi before. Both the duration and the depth or quality of a manager's past overseas experiences influence training needs. An expatriate will find the Indonesian culture less tough if he or she has lived there before than will the candidate who has not. Likewise, an expatriate who had frequent, in-depth interactions with Indonesians during a 3-year stay will find the cultural toughness less of an obstacle during a later visit than will the expatriate who had infrequent, superficial contact with Indonesians during a similar 3-year assignment. Because it is dangerous to assume that superficial experience will lead to positive results, both the quantity and the quality of a candidate's experience should be examined before deciding on the training required. Although this point may seem obvious, we have seen far too many cases like the following one.

Recently the general manager for a joint venture in Japan between a large U.S. telecommunications company and a Japanese company was selected in part because of his 3-year experience in Japan as a U.S. naval officer. Unfortunately, that

> **FROM THE FRONT LINES:**
>
> **"Training Women for Positions in Culturally Tough Environments"**
> by James Pilarski, Senior Vice President,
> International Human Resources, Marriott Corporation
>
> Many companies discriminate against women by not selecting them for international assignments (Stroh & Caligiuri, 1998). The assumption is that cross-cultural adaptation is more difficult for women than for men. This is definitely not the position we take at Marriott International. We actively seek women candidates for international assignments and have found our women international assignees to be as competent and able to adjust to new cultures as the men we send abroad. Katie Koehler is a perfect example.
>
> A few years ago, Marriott International sent Katie to Mexico City, where she was the director of human resources for our hotel there. Because of the many cultural differences in male–female relations, American women often find adjusting to Mexican culture and to other cultures in Latin America especially challenging. One of the best strategies companies can use to help their people adjust to culturally tough environments is to provide opportunities in predeparture training for employees to anticipate potential cultural problems and practice dealing with them. Katie was obviously well prepared when a problem arose.
>
> One day at a breakfast meeting, a top union leader told a sexist joke in Katie's presence. Prior to the assignment, Katie knew that she would probably be tested in this way. In fact, she had already prepared a response to the sexist joke. She gave the union leader a stern look and raised her eyebrow to subtly signal that she did not think the joke was funny but that she would not make a scene that might embarrass the union leader (and potentially ruin their relationship). Her technique obviously worked, for Katie and the union leader developed a good relationship.
>
> At Marriott International, we think that both men and women need to prepare for international assignments. As Katie quickly discovered, one of the best ways to manage culturally difficult situations is to anticipate a culturally appropriate response.

the individual had little direct interaction with Japanese during that time and lived on base were not carefully considered. The manager's previous limited experience did little to help prepare him for all the cross-cultural management intricacies of running the joint venture. He was sent without any training. Because key company officials did not think about the quality of this individual's previous experience, they were surprised when he encountered significant cross-cultural adjustment and management difficulties.

Communication Toughness

Another issue that should determine the rigor of training is how much the global manager will be expected to interact with the local populace. The more interaction is required, the higher the level of communication toughness. For example, an oil rig expert sent to Saudi Arabia may only rarely have to speak to a Saudi, either on or off the job, whereas a marketing manager in Peru may be in constant contact with local clients, advertising agencies, and people in the media. Obviously, the manager in the former situation does not require nearly as rigorous cross-cultural training as the manager in the latter situation does.

The extent of communication toughness in the global assignment can be determined by examining the extent to which the overseas job requires communicating with local nationals. Responses to the following questions will help you understand the degree to which such interaction will be necessary:

1. Are the rules and norms for communicating very different from those in the home country, or are they quite similar?
2. Will the manager have to communicate frequently or infrequently with local workers?
3. Does the expatriate candidate speak the host language? Do host nationals speak the expatriate's native language? If not, how difficult is the host language to learn?
4. Will the manager have to communicate mainly in one direction (e.g., giving orders, giving presentations, delegating, etc.), or will the nature of the job require intensive two-way communication with host nationals (e.g., teamwork, business negotiations, etc.)?
5. What will the main form of communication be with host nationals? Face-to-face communication or technical forms of communication (e-mails, memos, etc.)?
6. How long will the expatriate be working in the country? Six months? One year? Three years?
7. What will be the main type of interaction the expatriate will have to engage in with host nationals? Formal (figurehead-type functions, giving orders) or informal (influencing clients one-on-one, developing relationships with government officials)?

As the intensity of the interactions increases, the need for rigor in the training program increases proportionately. For example, if the communication will be frequent, two-way, face-to-face, and informal, the training should be more rigorous than in the opposite situation. In assessing communication toughness, it is important to recognize that even if the international assignee does not have a high level of interaction on the job, he or she may still have to undertake high levels of communication with locals outside the workplace. For example, a manager in Korea (a culture known for its high degree of communication toughness) may not have a great deal of daily interaction with local nationals but may have high levels of interaction as he or she manages the personal necessities of life. To ensure that managers and their families adjust to life in Korea, predeparture training should address communication both inside and outside the business arena.

Job Toughness

Many global managers are promoted when they are sent overseas. The promotion often means a job challenge because the manager is working in a new area and has more responsibilities, more autonomy, and new challenges. The tougher the tasks of the new job, the more assistance the manager will need through rigorous predeparture training. The elements of job toughness can be discerned in the answers to the following questions:

1. Are the performance standards the same?
2. Is the degree of personal involvement in the work unit the same?
3. Is the task the same or quite different?
4. Are the administrative procedures similar?
5. Are the resource limitations the same?
6. Are the legal restrictions similar?
7. Are the technological aspects of the job similar?
8. Is the freedom to decide how the work gets done the same?
9. Are the choices about what work gets delegated similar?
10. Is the freedom to decide who does which tasks the same?

An examination of the answers to these questions should enable a rough estimate to be made regarding the job toughness of the overseas assignment. If the job toughness is moderate to high, the manager will need job-specific training as well as cultural training in how jobs get done and how people are used to being managed in the country of assignment. If the demands of the job are quite different from the job the manager has had at home, if the constraints are greater, and if there is less freedom, then the job toughness will be greater as will the required level of training rigor.

Applying the Principles

Let us return to Mel, the manager at Recor. His predicament is very common across industries and in firms of all sizes. Remember that Mel was trying to put together a training program for 18 engineers who were being sent to Japan. Let's look at his situation and apply the principles discussed in this chapter to the design of an effective cross-cultural training program.

Analysis and Application

To do this, we will take a look at each of the three major determinants of training rigor—cultural toughness, communication toughness, and job toughness.

Cultural Toughness. Looking at the list of seven regions, Mel quickly determined that the Far East was the third-most difficult in both business practices and culture for Westerners to adjust to. After speaking with local university professors and local businesspeople who had had dealings with the Japanese and after consulting a few recommended books, Mel learned that the Japanese accept power differences in organizations and are more influenced by matters of hierarchy and status than Americans are. He also learned that the Japanese are less risk-oriented than Americans, are more comfortable working in groups, accept traditional sex-role differences, and have a stronger work ethic than Americans. The language is difficult to learn, and few Japanese speak fluent English.

Many of Mel's people had vacationed once or twice overseas (mainly in Europe and the Caribbean), but none had lived or worked abroad or in Japan. Thus, their previous experience would not reduce the cultural toughness or the need for rigorous cross-cultural training. Mel expected his people to experience serious culture shock on their arrival in Japan and that it would take some time for them to adjust to living and working there.

Communication Toughness. At first, Mel thought that his people would have a fairly low degree of interaction with host nationals. On reflection, however, Mel realized that the group-oriented nature of the Japanese organization and the practice of group decision making increased the likelihood that all of Recor's people would be interacting frequently with the Japanese at work. Mel surmised that his team of engineers would be working in a business culture with different communication rules where they would need to interact frequently with the Japanese, communicate with people speaking a difficult foreign language, and be expected to carry out tasks that would require two-way, face-to-face, informal communication. Communication toughness would be high.

Job Toughness. By contrast, it appeared on the surface that the job toughness in Japan would not be high. All the engineers had considerable professional experience; they did not need to learn any new technical skills. Nevertheless, there was a good chance that performance standards, tasks involving training or working with the Japanese, the ways in which decisions are made, and the bureaucratic procedures that have to be followed would all be somewhat different.

Closer examination of the various aspects of the job suggested that job toughness would probably be moderate rather than low.

So far, Mel's analysis focused on the engineers. What about their families? Several people had emphasized to Mel that family issues should not be overlooked; overseas assignments frequently get aborted because the spouse and the family are unable to adjust to the new culture. In general, family members react to cultural toughness as much as managers do, so the family needs predeparture training too. Children under 13 seem to have less difficulty adjusting to new cultures than teenagers do, but ideally, the cross-cultural training teenagers and spouses receive should be as rigorous as the training given managers. Nonworking spouses find the adjustment to cultures that are high in communication toughness especially difficult: The inability to communicate both verbally and nonverbally can lead to depression, alienation, and loneliness. Telling people to "snap out of it" is simply not enough. Mel has to weigh all these issues before deciding on the training design. His budget may be limited, but if at all possible, he needs to ensure that the engineers' spouses and other relevant family members are included in the training program.

What conclusions can Mel reach from his analysis of the culture in which his team will be working? His findings can be summarized as follows:

1. Cultural toughness = high.
2. Communication toughness = high.
3. Job toughness = moderate.

Framework for Developing Cross-Cultural Training

Figure 4.2 illustrates a framework for selecting a training program (Black & Mendenhall, 1989). The reasoning behind it is straightforward: the greater the cultural toughness, communication toughness, and job toughness, the greater the need for rigorous cross-cultural training. Mel heeded experts' views that the training should take place in two stages (predeparture stage and in-country stage) and not be limited to predeparture training only (Feldman & Bolino, 1999; Mendenhall & Stahl, 2000; Selmer, Torbiörn, & de Leon, 1998; Suutari & Burch, 2001).

Stage 1: Predeparture Training

Mel realized that this group would need at least training at the high end of moderate. He had no in-house expertise and could get the team members together for only 3 to 5 days of training. After examining several programs offered by various consulting firms, Mel found only one firm that offered a program that had content that he considered rigorous. Mel encouraged the spouses to attend the training as well, and all did but two.

Low Training Rigor (Duration = 4–20 Hours)	Moderate Training Rigor (Duration = 20–60 Hours)	High Training Rigor (Duration = 60–180 Hours)
Lectures Films Books Area Briefings	Methods in previous box, plus: Role Plays Cases Assimilators Survival-level Language	Methods in previous box, plus: Assessment Centers Simulations Field Trips In-depth Language

FIG. 4.2. Determining cross-cultural training vigor.

Mel and the members of the consulting firm agreed that the focus of the training would be the development of cross-cultural interaction skills. Among the specific methods used were role plays, short simulations, culture assimilators, and case studies. Mel arranged for the engineers and their families to take lessons in the Japanese language after arriving in Japan. In addition, Mel gave them a list of books on Japanese culture, compiled after his assistant consulted with Japanese experts at the local university.

During his research on his problem, another issue came to Mel's attention regarding the rigor of predeparture training—namely, that one-size-fits-all training may not be as effective as tailoring the training to the specific context of the international assignment (Vance & Paik, 2002). The predeparture training offered by the consulting firm was rigorous in nature but contained general content. That is, it rigorously delved into general Japanese culture norms. New research is beginning to find that in addition to general cultural norms, predeparture training should be custom designed around the unique contextual and cultural issues associated with the circumstances the international assignees will face—in this case, a joint venture in which engineers would be working closely with each other. Mel asked the consulting firm to get input from the Japanese engineers regarding their expectations, concerns, goals, and needs regarding working with the U.S. engineers. That way, Mel could ensure that the predeparture training would address both general and joint venture-specific cultural issues.

Mel realized that although predeparture training was necessary to help the engineers and their families start off on a good foot cross-culturally, they would need more training during their sojourn overseas. His research indicated that Japanese business and social culture were complex and often fundamentally different from U.S. culture. Thus, Mel began a search for experts who could provide his people with training throughout the joint venture project.

Stage 2: In-Country Training

Mel's concern about helping his people develop while overseas brings up an important consideration that is overlooked by virtually all consulting firms that do cross-cultural training. Learning about the new culture is useful before leaving on a global assignment, but truly effective training requires rigorous, in-depth, cross-cultural training after global managers are "in-country" (Feldman & Bolino, 1999; Mendenhall & Stahl, 2000; Selmer, 2001; Selmer et al., 1998).

To understand why this is the case, refer to Fig. 4.3 and think about Mel's group of engineers before and after they arrive in Japan. Before they leave their home country, it is more difficult for them to imagine what the trainers are trying to explain. For example, it is one thing to imagine what it is like to have a subordinate in Japan say, "Yes, I understand what you want me to do" and then do nothing because in fact, he did not understand. It is another thing to have such an interaction and then learn about the cultural reasons behind it. In-country training has several advantages: (a) trainees are more highly motivated; (b) they have a higher level of baseline experience with the local culture as a foundation for learning deeper cultural values, norms, and ideas; (c) the trainees can immediately apply what they learn; and (d) the environment itself makes the training content real.

Predeparture training should focus mostly on basic, day-to-day, survival-level concerns. These are the issues that the assignee and family will have to deal with as soon as they step off the plane. Predeparture training should also include some of the deeper aspects of the culture but should not attempt to cover every segment at the deepest levels. In addition to the reasons we have already mentioned, without some actual experience in the culture, many managers will simply find it hard to believe that "things could really be that different."

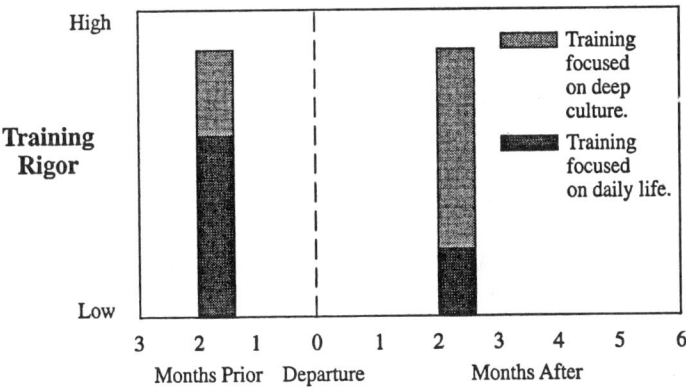

FIG. 4.3. Cross-cultural training: content and timing.

The bulk of in-depth culture training should take place after the assignee has been in the country for at least a month but not more than 6 months (see Fig. 4.3). The reason to wait at least a month is simple. During the first month after arrival, the assignee and family are so involved in all the logistics of getting settled that they have precious little energy or mental capacity to absorb in-depth cultural training. The rationale for not waiting more than about 6 months is equally straightforward. After several months in the country, the assignee and family will begin to form judgments and conclusions about the culture—what the norms are, what motivates people, and so on. Even if they are wrong, changing their minds at that point is difficult at best, impossible at worst.

There is no way a traditional training program can cover all the possible permutations of cross-cultural interactions that Mel's engineers will face. Thus, the key to in-country training is to expand one's view of what a training program should cover. It has been argued that the most effective approach is to retain a local expert who is on call to address expatriates' specific concerns and needs while they are overseas. It is important that they have someone they can talk to one-on-one to help them during the process of cross-cultural adjustment. During the overseas assignment, a coach—someone who is well versed in Japanese business and social cultures—will be able to better facilitate learning among Mel's expatriates than some canned training program. Of course, if the coach perceives that a majority of the expatriates are experiencing similar cross-cultural challenges, he or she could hold a program to address specific issues with the group as a whole. Flexibility is the key to management development during an overseas assignment (Mendenhall & Stahl, 2000).

Linking Corporate Orientation to Training

In chapter 1 (this volume), we discussed the importance to firms that are "going global" of maintaining a fit between their globalization stage, their competitive strategy, and their people management. Each of the five functions of people management (in this case, the training function) must support the others and not detract from them.

Internal Fit

Spending a lot of time and money carefully searching for key people to send to specific countries can turn out to be a waste of an investment if the company uses an invalid, generic, ineffective training program. Likewise, offering no training at all is certain to inhibit the effectiveness of the selection process. Thus, the training function is an important link between good selection and good management development in the process of establishing internal fit.

A companywide, canned, cross-cultural training program may not support the function of management development either. Imposing a one-dimensional approach on all international assignees, with no regard to their current strengths and weaknesses, to the unique situations they will find themselves in overseas or to the kinds of skills they need to learn will simply create frustration, unmet expectations, and poor management on the part of the global managers when they are overseas.

External Fit

Cross-cultural training, like all the other functions of people management, must be designed with the firm's stage of globalization in mind. Consider two global managers, both assigned to Nigeria. One works for a firm at the export stage of globalization, whereas the other works for a multinational corporation. Even if both jobs appear similar in cultural toughness, communication toughness, and job toughness, the globalization stages of the respective firms should also influence how training is designed. These relations are illustrated in Fig. 4.4.

The matrix in Fig. 4.4 overarches what we have discussed to this point. The assumption in designing this matrix was that cultural toughness, communication toughness, and job toughness were always identical. This was done to illustrate

Export Stage	MDC Stage
Degree of Rigor: Low to Moderate	*Degree of Rigor:* Moderate to High
Content: Emphasis should be on interpersonal skills, local culture, customer values, and business behavior.	*Content:* Emphasis should be on interpersonal skills, local culture, technology transfer, stress management, and business practices and laws.
Host-Country Nationals: Low to Moderate training of host nationals to understand parent country products and policies.	*Host-Country Nationals:* Low to Moderate training of host nationals; primarily focusing on production/service procedures.
MNC Stage	**Global Stage**
Degree of Rigor: High-Moderate to High	*Degree of Rigor:* High
Content: Emphasis should be on interpersonal skills, two-way technology transfer, corporate value transfer, international strategy, stress management, local culture, and business practices.	*Content:* Emphasis should be on global corporate operations/systems, corporate culture transfer, customers, global competitors, and international strategy.
Host-Country Nationals: Moderate to High training of host nationals in technical areas, product/service systems, and corporate culture.	*Host-Country Nationals:* High training of host nationals in global organization production/efficiency systems, corporate culture, business systems, and global conduct policies.

FIG. 4.4. Stage of globalization and training design issues. MDC = multidomestic corporations; MNC = multinational corporations.

how the globalization stage should influence the design of cross-cultural training. In general, the more a firm is moving out of the export stage of globalization, the more rigorous the training should be. Breadth of content also increases as the firm moves from the export stage to the global stage; in other words, more rigorous training is needed. As firms move from the export and multidomestic stages, global managers also need to be able to socialize host-country managers into the firm's corporate culture and other firm-specific practices; this managerial responsibility requires rigorous training.

Consider, for example, some of the training issues that arise at a company like Nestle as a result of its globalization orientation. Nestle is a firm that responds well to local markets and that has relatively low levels of coordination of activities across borders. In some ways, Nestle could be classified as a multidomestic organization. In-depth knowledge of each culture and market where the firm operates is critical because production and product innovations take place in a variety of countries. Although local nationals would be more expert than expatriates in local cultural and market subtleties, Nestle global managers retain control of these operations. Consequently, the number of managers Nestle sends overseas is higher than if Nestle were simply exporting. In this context, the rigor of the training offered to the global managers whom Nestle deploys ranges from moderate to high (see Fig. 4.1). The primary reason for this level of training is that to be effective, a global manager must have specific knowledge of the target foreign market and culture. Thus, the training needs to emphasize skills related to local business practices and cross-cultural communication, stress management, and so on. Training that included tactics for socializing local nationals to the Nestle philosophy, implementation of worldwide corporate systems, and other firm-homogenizing practices would not be particularly useful because such practices do not fit the globalization orientation of the firm as a whole.

Whether a firm has a pattern of globalization that requires many or few managers to be in positions around the world, those managers need to receive training appropriate to the contexts of their assignments. To maintain external fit, the training function must be flexible enough to deal with all potential contexts that derive from a firm's globalization patterns. A rigid, mechanical training philosophy will not benefit global managers to the greatest extent possible. Following the framework presented in this chapter, wise training decisions can be made despite constraints in the organization and the environment.

PART
III

DURING THE ASSIGNMENT

CHAPTER
5

Adjusting: Developing New Mental Road Maps and Cultural Skills

In chapter 2 (this volume), we addressed people's reactions as they struggle to adapt to living and working in an unfamiliar culture. In chapters 3 and 4 (this volume), we discussed whom to select and how to train people for international assignments. This chapter focuses on some of the factors affecting cross-cultural adjustment among expatriates.

You may recall that in chapter 2 (this volume), we pointed out that a culture is characterized by a set of rules, values, and assumptions. These common elements of the culture are passed from one generation to another and influence how people in the culture conduct business, how they interact with their friends and associates, and how they behave in restaurants and other public settings. As we also mentioned, the most powerful components of a culture are invisible; consequently, they are hard to recognize, understand, and adjust to. We also mentioned that being immersed in a new culture disrupts familiar and established routines and threatens one's sense of identity and self-image. Culture shock, which manifests as feelings of anger, frustration, and depression, is a common response as one struggles to make sense of the culture's rules, values, and assumptions and function effectively. Many people return home early as a way to escape these feelings and the situations that cause them. Whether out of fear, determination, or both, most people continue to live and work in the foreign country, struggling to adjust.

Two interrelated components are essential to the process of cross-cultural adjustment. First, the international assignee must gain *predictive control* (Bell & Staw, 1989). Predictive control begins when an individual starts to understand the logic behind a culture's values and how those values are applied (Bird, Osland, Mendenhall, & Schneider, 1999). With this understanding, the individual can create a new set of mental road maps and traffic rules that enable him or her to predict what behaviors are expected in specific situations, how people will probably respond, what behaviors are not appropriate, and so on. This is a critical step in gaining a sense of control over the new environment and thus essential to the process of cultural adjustment.

Second, the expatriate must gain *anticipatory adjustment* (Selmer, 2002). This involves figuring out the reactions one's behavior is likely to elicit in the culture and mastering the behavior necessary to produce desired rewards. For example, it is one thing to know that workers in Latin America may be reluctant to volunteer their opinions out of deference to a senior manager's authority but a lot more difficult to adjust one's management style to accommodate when a project requires subordinates' input.

It is important to keep in mind that adjustment to a new culture does not require total and permanent transformation. It is possible to adjust effectively without embracing all of a culture's assumptions, values, and behaviors. Adjustment does not require rejecting one's native culture, either. However, effective adjustment does require developing an appreciation of the local culture. Today's global managers need to be bicultural. They need to speak the local language, be well integrated into both the local culture and the community of expatriates, and understand the host culture as well as their own (Kedia & Mukherji, 1999).

DIMENSIONS OF CROSS-CULTURAL ADJUSTMENT

Most research on cross-cultural adjustment has focused on American men working overseas. Recently, however, work has been done on European managers assigned to China and on Australians assigned to Thailand (Clegg & Gray, 2002; Selmer, 2002). Much more work has been undertaken on female expatriates as well (Stroh et al., 1998). Consequently, although the major influences on cross-cultural adjustment continue to be based on the experiences of American males, we also discuss the extent to which these factors are true for women and for non-Americans living abroad.

When researchers first began examining adjustment among expatriates, they focused on their adjustment to the more obvious aspects of their new cultures, such as the food, weather, daily customs, and so on. More recently, the authors of this book have discovered three related aspects or dimensions of adjustment that are less visible but equally or more important to overall adjustment to a new culture: adjustment to the job, to interacting with host-country nationals, and to the general nonwork environment. Other researchers have also found strong evidence that these three dimensions of cross-cultural adjustment apply to both U.S. and non-U.S. managers, male and female (Shaffer, Harrison, & Gilley, 1999).

These three dimensions are represented in the far right-hand box of Fig. 5.1.

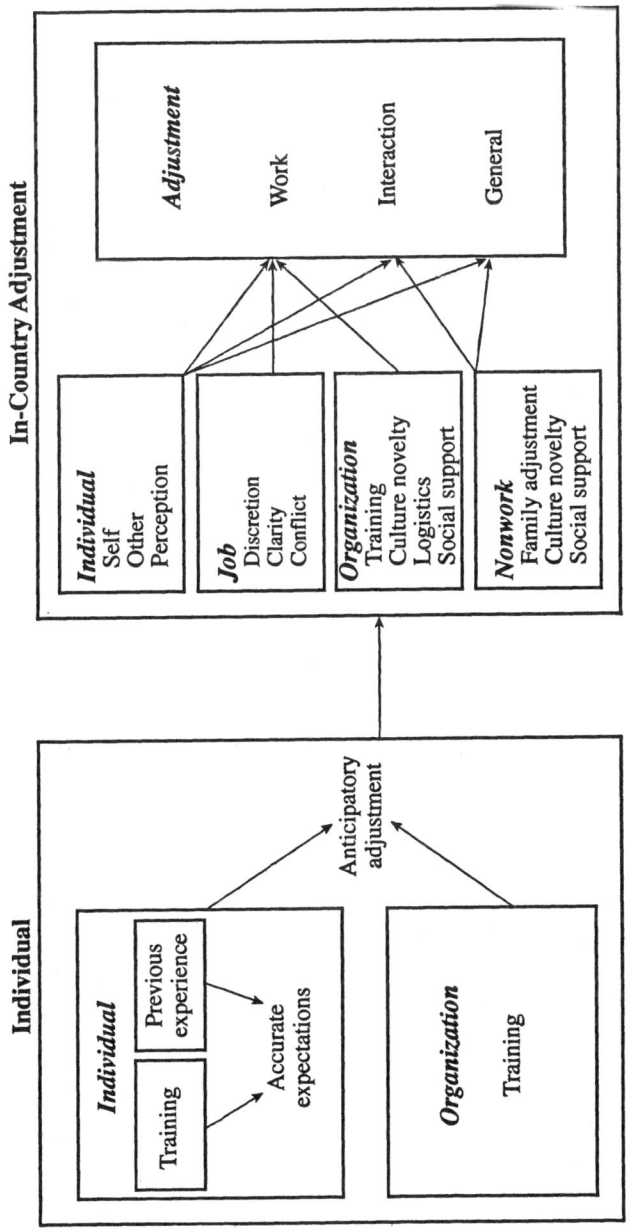

FIG. 5.1. Dimensions of cross-cultural adjustment.

Adjustment to the Job

Most expatriate managers find that of the three dimensions of adjustment, adjusting to work is the easiest primarily because there are usually similarities in procedures, policies, and requirements of the job both at home and in the foreign operation. Although adjustment to the job may be the easiest adjustment to make, adjustment to work can still be difficult.

Aspects of the foreign operation's corporate culture as well as its national culture are often dramatically different from those back home and therefore affect a manager's responsibilities and the performance of tasks. Consequently, although a manager from Los Angeles may be expected to perform basically the same tasks in Beijing, elements of the foreign operation and the host-country culture may make it necessary to perform tasks in a slightly or even a dramatically different manner to achieve similar results. Many day-to-day elements of decision making, for example, vary greatly across cultures. In the United States, for example, delegating decision making and oversight to a subordinate might be a sign of respect. However, in many countries in Latin America and southern Europe, a subordinate might be totally unprepared to handle this additional responsibility. This could cause breakdowns in communication between the manager and the subordinate and potentially delays and problems in completing the work.

Adjustment to work may also be affected by gender relations and compensation levels in the host country. There is some evidence, for instance, that adjustment to work is more difficult for women in countries where they are underrepresented in the workforce (Caligiuri & Tung, 1999; Caligiuri, Joshi, & Lazarova, 1999).

Adjustment to Interacting With Locals

Regardless of the expatriate manager's nationality, interacting with host-country nationals is generally the most difficult of the three adjustment dimensions primarily because differences in values and assumptions are most likely to affect these interactions. Time spent with other expatriates prior to the assignment and the extent to which the new culture is different from one's home culture significantly affect adjustment along this dimension (Selmer, 2002).

In the United Kingdom, for example, a plant manager might simply interview workers to find out their needs for and expected uses of a proposed piece of equipment. By contrast, in Mexico, workers may expect to be told how they are to use the new equipment and may be confused when someone asks for their input. As a result of cultural differences, a British expatriate may find that asking Mexican workers how they would use the equipment will generate information that is incomplete and inaccurate.

Adjustment to Nonwork Environment

Adjustment to the nonwork environment has been the focus of much research on cross-cultural adjustment. In general, international assignees have an easier time adjusting to the general environment than to interacting with host nationals but have more difficulty adjusting to the general environment than to the job. Recent research, however, suggests that adjustment to the nonwork environment may affect adjustment to the job and subsequently adjustment to interacting with host-country locals (Takeuchi, Yun, & Russell, 2002).

Included in adjustment to the nonwork environment are adjustment to food, transportation, entertainment, health care, and other nonwork issues. Several factors affect overall nonwork adjustment, including international experience, time spent with other expatriates, and the novelty of the new culture (Aycan, 1997).

Case 13: Fred Bailey—An Innocent Abroad

One of the easiest ways to understand the three dimensions of adjustment is to examine a real (but disguised) case of an expatriate manager. Fred Bailey and his family had recently been assigned to the Tokyo office of a Boston-based, U.S. consulting firm.

Fred Bailey gazed out the window of his 24th-floor office at the tranquil beauty of the Imperial Palace amid the hustle and bustle of downtown Tokyo. It had been only 6 months since Fred arrived with his wife and two children for his 3-year assignment as director of Kline & Associates' Tokyo office. (Kline & Associates was a large multinational consulting firm in Boston with offices in 19 countries.)

Fred was trying to decide whether he should simply pack up and tell the home office that he was leaving or try somehow to convince his wife (and himself) that they should stay and finish the assignment. Fred had thought they all were excited to begin with, so it was a mystery to him how things had reached this point. As he reflected back on a variety of incidents, one stood out as particularly frustrating.

Not long after his arrival in Japan, Fred had a meeting with representatives of a top Japanese multinational firm concerning a potentially large contract. Those present included Fred; the lead American consultant for the potential contract, Ralph Webster; and one of the senior Japanese associate consultants, Kenichi Kurokawa, who spoke perfect English. The Japanese team had four members: the vice president for administration, the director of international personnel, and two staff specialists. After some awkward handshakes and bows, Fred said that he knew the Japanese gentlemen were busy and he did not want to waste their time, so he would get right to the point. Fred then had Ralph Webster lay out the firm's proposal for the project and what the project would cost. After the presentation, Fred asked the Japanese team for its reaction to the proposal. The Japanese team did not respond immediately, so Fred launched into his summary version of the proposal, thinking that the translation might have been insufficient. Again, although he asked several direct questions, the Japanese team offered only vague responses.

After 5 months, a contract between the firm had yet to be signed. "I can never seem to get a direct response from the Japanese," Fred complained.

Fred decided that the reason little progress had been made was that he and his group did not know enough about the client to package the proposal in a way that was appealing. He called in Ralph Webster and asked him to develop a report on the client so that the proposal could be reevaluated and changed as necessary. Jointly, Fred and Ralph decided that one of the more promising young Japanese research associates, Tashiro Watanabe, would be the best person to take the lead on this report. To impress on Tashiro the importance of this task and the great potential they saw in him, they decided to have the young Japanese associate meet with them. In the meeting, Fred had Ralph describe the nature and importance of the task. Fred then leaned forward in his chair and said, "You can see that this assignment is important and that we are placing a lot of confidence in you by giving it to you. We need the report by this time next week so we can revise and present our proposal again. Can you do it?"

After a pregnant pause, Tashiro responded hesitantly. "I'm not sure what to say."

At that point Fred smiled, got up from his chair, walked over to the young Japanese associate, extended his hand, and said, "Hey, there's nothing to say. We're just giving you the opportunity you deserve."

The day before the report was due, Fred asked Ralph how it was coming. Ralph said that because he had heard nothing from Tashiro, he assumed everything was under control but that he would double check. Ralph later ran into one of the American research associates, John Maynard. Ralph knew that John had been hired because he was fluent in Japanese and that, unlike the other Americans, John often went out after work with some of the Japanese research associates including Tashiro. Ralph asked John if he knew how Tashiro was coming along on the report.

John recounted that the night before at the office, Tashiro had asked him if Americans fire employees for being late with reports. John, sensing that the question was not hypothetical, asked Tashiro why he wanted to know. Tashiro did not respond immediately. Because it was 8:30 p.m., John suggested they go out for a drink. At first Tashiro resisted, but John assured him that they would grab a drink at a nearby bar and come right back. At the bar, John got Tashiro to open up.

Tashiro explained the nature of the report he had been requested to produce. He continued to explain that although he had worked late every night, it was impossible to complete the report in a week and that he doubted from the beginning that he could complete it that soon. Furthermore, Kenichi Kurokawa, who was 4 years senior to Tashiro, had originally been involved in the project but apparently nothing had been said to him about Tashiro's assignment.

At this point, Ralph asked John, "Why the hell didn't he say something in the first place?" Ralph did not wait to hear whether John had an answer to his question. He headed straight to Tashiro's desk.

The incident got worse. Ralph chewed Tashiro out and then went to Fred, explaining that the report would not be ready and that Tashiro had not thought it could be from the start.

"Then why didn't he say something?," Fred asked. No one had any answers, and the whole thing left everyone more suspicious and uncomfortable.

The problem centered around different communication styles and assumptions. Fred's operating assumption was that if you wanted someone to understand your idea, opinion, and so forth, it was your responsibility as the speaker, not the listener, to clearly and explicitly communicate what you had to say. In Japan, the cultural assumption regarding communication is quite different. Tashiro assumed that the listener (Fred in this case) had significant responsibility for hearing what was not said. Tashiro assumed that Fred would hear Tashiro's worries even if they were implicit rather than explicit. When asked why he hadn't voiced his concerns about the feasibility of completing the project on time, Tashiro responded, "I did," meaning that he expressed his concerns without explicitly stating them. A good Japanese manager would have picked up on these implicit, unvoiced concerns. In fact, it would have been the Japanese manager's responsibility to do so.

Fred found himself frustrated in interactions with Japanese customers and coworkers because the two cultures hold very different assumptions about communication. In general, Americans expect explicit communication and place most of the responsibility on the speaker. In contrast, Japanese value implicit communication and place responsibility for effective communication on both the speaker and the listener.

An encounter the Baileys had at a Japanese resort illustrates the potential challenges of adjusting to the general environment. After a few months into the assignment, the Baileys were looking forward to a long weekend vacation, and everyone in the office had advised Fred to stay at a famous hot springs north of Tokyo. Fred, Jenny, and their children arrived and were proceeding to the front desk to check in when an elderly Japanese man started making gestures with his hands and speaking rapidly in Japanese. Fred and Jenny stopped in time, but the two girls had stepped up to the main floor with their shoes on before the man could bring them slippers. Once they all had their multicolored slippers on and their shoes securely stored in lockers, the Baileys proceeded to check in.

This accomplished, two elderly Japanese women escorted the Baileys to their room. After walking down a long hall, the women stopped in front of a wooden door, slid it back, and waited for the Baileys to enter. The hall door opened into a small entry room, about 3 ft. wide and 5 ft. long. The Japanese women indicated that they should all remove their slippers, but, once again, the girls had opened a door and stepped into the main room. After removing their slippers, Fred and Jenny gazed in amazement at the big, empty room; there were no beds and no furniture. The Japanese women showed the Baileys the closet in which the futons, or sleeping mattresses, and blankets were stored, and they rattled off several comments in Japanese. As the Japanese women prepared to leave, Fred tried to tip them. The women covered their mouths and waved their right hands back and forth in front of their faces.

Later, Fred decided to take a soak in one of the hot springs. He went downstairs to the locker room and took off his clothes. Two doors led out of the locker room, and the signs above them were in Japanese. Fred decided to take the door on the right and entered a large room. On the right side was a line of short wooden stools and small buckets lined up in front of water faucets. At one of the faucets, a Japanese man was seated, vigorously washing his hair. In the center of the room was a large tub about the size of a swimming pool. At about the same time Fred noticed this, he also heard giggles coming from three young women sitting in the water on the left side of the tub. Fred had stumbled into the "co-ed" section of the hot spring. He quickly retreated, got dressed, and returned to his room.

Later that evening, Fred and his family decided to have a fish dinner. After a few strange appetizers, a big platter of fish was brought in. As the fish was placed on the table in front of the Baileys, Fred noticed a movement. As he looked closer, Fred could see the gills of the fish opening and closing in a futile attempt to breathe. The body of the fish had been masterfully sliced into little fillets that lay all along its torso. Christine shrieked, "Daddy, it's still alive! They've sliced it open, but it's still alive!" Suddenly no one was hungry.

What the Baileys failed to appreciate was that the way the fish was prepared—quickly "gutting" and skillfully slicing it into sashimi (raw fish fillets) and serving it before the fish's reflexes stopped it from opening and closing its gills—demonstrated absolute freshness. This was a skill and level of service very few resorts could offer.

CATEGORIZATION OF FACTORS INFLUENCING ADJUSTMENT

One of the important reasons for distinguishing the dimensions of cross-cultural adjustment is that not all factors affecting adjustment are related equally or even at all to the three dimensions of adjustment discussed previously. For example, a difference in procedures or policy may influence work adjustment more than one's ability to interact with the locals and have no influence at all on adjustment to the general nonwork environment.

We have organized the various factors affecting adjustment into two categories—those that influence adjustment before departure and those that influence adjustment after arrival. These factors are represented in Fig. 5.1 by two separate boxes.

We have further divided the factors that influence adjustment after arrival into four separate categories: factors related to the individual, to the job, to the organization, and to nonwork issues. In the following section, we discuss, for each of these categories, specific factors that researchers have found are important and detail the adjustment dimensions on which the factors tend to have the strongest influence. Figure 5.1 illustrates various relations among these influences.

Factors Affecting Adjustment Before Departure

One of the realizations that has emerged from recent research on cross-cultural adjustment is that the effective global manager is far more than merely someone with corporate experience and technical expertise. Global managers show high levels of sensitivity to cultural diversity and are willing to make the sacrifices and commitment necessary to achieve acceptance in the host country (Downes & Thomas, 1999; Kedia & Mukherji, 1999). Candidates who have absolutely no desire to accept an international assignment are the least likely to make the adjustments in their mindset in advance of the assignment (anticipatory adjustment) and are the least likely to adjust after they are relocated (Selmer, 2001b).

The process of anticipatory adjustment is primarily psychological. The manager going on an international assignment begins to adjust his or her mental mindset, including maps and rules, by developing *anticipatory expectations.* For example, if individuals know in advance that the British drive on the left side of the road, they can make mental adjustments before actually traveling to Great Britain. Likewise, if a manager knows that the Japanese never wear their shoes on *tatami* (mats), the manager can mentally adjust before receiving an invitation to visit a Japanese colleague at home. Generally, three aspects of anticipatory adjustment make in-country adjustment easier: (a) making a large number of mental adjustments in advance, (b) focusing the process of anticipatory adjustment on important aspects of the new culture, and (c) refining the adjustments so that they are accurate. Significant research has been conducted on realistic expectations about future jobs and assignments and the impact of these expectations on adjustment (Stroh et al., 1998).

Individual Factors. Research has found that two specific factors facilitate the formation of accurate expectations and mental adjustments. The first factor is one we have already discussed in some detail: cross-cultural training. (In particular, issues concerning the content and rigor of training were examined in chap. 4, this volume.) Training can be categorized as an individual or an organizational factor because it is something individuals can provide for themselves or can receive through their firms.

The second individual factor that can help someone develop accurate expectations is previous overseas experience. Many writers have speculated that having lived in a foreign country has a positive impact on the formation of accurate expectations about another foreign experience. The assumption is that learning how to live and work in Italy is applicable to learning how to live and work in China. Research suggests, however, that although previous experience can be positively related to all three forms of adjustment, just because someone has lived in one country does not necessarily mean that adjusting to a new culture will be easier. Previous overseas experience can have a positive impact on anticipatory adjustment, but because various aspects of living and working in one foreign country are not always applicable to another, the positive impact is moderate at best (Selmer, 2000, 2002).

The impact of previous international experience is strongest if the last international experience was relatively recent, if there was a relatively high degree of interaction with host-country nationals, and if the previous experience was in a country whose culture is similar to the one in the country where the person is being relocated. For example, consider the case of the French manager who lived in South Africa for 2 years before his current assignment there. His previous assignment was 15 years ago, and he lived in a neighborhood dominated by families of French diplomats. He really did not have to interact with the local people and culture at all. In his current assignment, managing a joint venture between a government ministry and his firm, he will be living in a neighborhood populated by native South Africans. Under these conditions, his previous experience will not have a significant impact on the formation of accurate predeparture expectations and therefore will not facilitate his successful adjustment or job performance. Because someone worked overseas many years ago is not an accurate predictor of success in an impending assignment.

Organizational Factors. Evidence suggests that predeparture cross-cultural training can have a positive impact on the development of cross-cultural skills, adjustment, and job performance (Aycan, 1997; Eschbach, Parker, & Stoeberl, 2001). However, in the weeks preceding relocation, most managers are preoccupied with basic logistical issues. Further, because most of them do not have recent experience in the countries to which they are being sent, their level of motivation and general capacity to relate to deep cultural information are often not ideal just before they are transferred. Consequently, research suggests that some (if not most) of the deep cross-cultural training should be provided a month or so after relocation. International managers who receive intense predeparture and postarrival cross-cultural training tend to experience less devastating culture shock than managers who do not receive such training (Eschbach et al., 2001).

> **FROM THE FRONT LINES:**
>
> **"Intel's Mentoring Program Helps International Employees"**
> by Kevin Gazzara, Program Manager,
> Managing Through People, Intel
>
> Most mentoring programs focus on individual career issues. In Intel's mentoring program, the emphasis is on sharing information, skills, and knowledge across organizations and leadership ranks. In the case of its global workforce, employees share Best Know Methods and ideas between multiple different geographic locations. Intel uses a "virtual factory" concept to help new organizations and factories ramp as quickly as possible while ensuring highest quality approaches.
>
> "Intel's mentoring movement," as it was coined in the March 2002 issue of *Fast Company Magazine*, is more structured than many such programs in other companies and industries. Employees who have agreed to be mentors complete an on-line questionnaire in which they list any of their two dozen top skills. Mentees indicate what they are interested in learning, and with that information, mentees are provided a list of mentors to choose from. Mentors and mentees may be in the same location, or they may be in locations at different ends of the country or the world. Partnership contracts and deliverables are used to ensure the program gets the results the mentors and mentees desire. Intel's mentoring program has been especially useful in tapping into employees' knowledge and diverse experience for its newer factories and emerging global markets.

Another predictor of adjustment, in addition to training, is whether an organization's international managers receive company assistance and support (Kraimer, Wayne, & Jaworksi, 2001). Predeparture, as well as postarrival, this support should be both logistical and emotional. When companies show genuine concern for the well-being of the employee and his or her family, the chances are much better that they will adjust well to their new culture (Stroh, 1997). Recently, mentoring programs have received growing attention as a way firms can assist in the adjustment process of global managers and their partners. The mentors, who are typically managers with fairly extensive in-country experience, socialize with the newcomers and share practical information both before and after the employees arrive in their new country (Feldman & Bolino, 1999; Harvey, Buckley, Novicevic, & Wiese, 1999; Mayrhofer & Scullion, 2002). "From the Front Lines" offers a glimpse at Intel's formal but somewhat unusual approach to mentoring.

Factors Affecting Adjustment After Arrival

In this section, we discuss factors that affect adjustment after arrival in the foreign country (postarrival or in-country factors); however, many of these factors require planning and design before individuals are sent overseas. (Steps that organizations can take to facilitate successful cross-cultural adjustment are discussed in more detail at the end of this chapter.)

Individual Factors. Individual factors that influence cross-cultural adjustment were discussed in chapter 3 (this volume), but they can be summarized in terms of three broad categories. The first category consists of *self-oriented factors*. These characteristics relate to whether people believe in themselves and have confidence in their ability to deal effectively with people from a culture different from their own as well as with new surroundings. This is not to say that arrogance and an inflated self-image are positive attributes when trying to adjust to a foreign culture. However, people who believe strongly in themselves tend to persevere even in the face of mistakes, ask questions about mistakes they make, learn from their mistakes, and do not make the same mistakes repeatedly. Thus, individuals with strong, healthy self-images tend to persist in adjusting their behaviors even when these adjustments are less than perfect and result in blows to their egos. The more they try to master new behavior, the more they have the opportunity to receive both positive and negative feedback about how they are doing and to refine their behavior until it is effective.

The second category of individual factors are *other-oriented* or *relational factors*. These characteristics are associated with the ability to meet new people, interact with them, and empathize with them. This orientation toward developing relationships is extremely important because host-country nationals are the best source for teaching foreigners how to successfully navigate in their country. Host-country nationals care less about how skillful you are but how sincere you are in wanting to understand their culture. That is almost more important than the skills. Host-country nationals sense the degree to which you are really working hard, the degree to which you want to connect with them.

The third group of individual factors are *perceptual factors*. These include an individual's ability to grasp and understand invisible maps and rules. People are not equal in this ability. Some people cannot comprehend what is not visible. Others are much better able to appreciate and understand the subtle determinants of people's behavior.

Research remains to be done on the relative strength of specific individual factors; however, studies suggest that G–A–P–S is an effective way to measure characteristics that have a positive influence on all three dimensions of cross-cultural adjustment (Aycan, 1997; Black, 1990; O'Hara & Johansen, 1994).

Job Factors. Job factors tend to have their strongest influence on work adjustment and tend to have little spillover effect on interaction or general nonwork adjustment.

The job factor that has the strongest impact on work adjustment is job discretion. *Job discretion* is defined as the amount of freedom individuals have in their jobs. A person who has a great deal of job discretion has flexibility in determining what work to do, when and how to do it, and whom to involve. People with greater job discretion are able to configure their work so they can use past successful behaviors and approaches more easily, which in turn facilitates job adjustment and performance (Alampay, Beehr, & Christiansen, in press).

Another aspect of the job that helps improve work adjustment is *job clarity* or extent to which what is expected of the individual is unambiguous. It is hard to adjust to something that is unclear or ambiguous. Consequently, job ambiguity generally hurts job adjustment and performance, whereas job clarity has a positive impact on both outcomes.

Yet another aspect of the job that influences work adjustment is the extent to which conflicting demands or expectations are made on the worker. This is called *role conflict,* not to be confused with job ambiguity. In the case of job ambiguity, what is expected is clear but different people have conflicting expectations of the manager. Role conflict generally hinders work adjustment.

Organizational Factors. Four specific organizational factors have a significant impact on adjustment to an international assignment. These factors tend to be related most strongly to work adjustment rather than to how well the expatriate adjusts to interacting with the locals or to the general nonwork environment.

The first of these organizational factors is postarrival cross-cultural training. Ideally, this training should focus on the specifics of the job, how to interact effectively with host nationals, and on the general environment. Training that is comprehensive yet country specific is most likely to have an impact on all three aspects of adjustment. Such training helps individuals gain the necessary mental maps and understand the rules of a culture; it also gives them opportunities to practice and develop the behaviors and skills necessary to operate effectively in the foreign culture.

Too often, firms simply use training as a means of conveying information. Understanding a culture's rules is a necessary but insufficient step in the process of cross-cultural adjustment and competence. To behave appropriately, people also need to understand the logic of the culture at a deeper level. (As discussed in chap. 4, this volume, developing these skills and behaviors requires rigorous training.)

The second organizational factor that can affect work adjustment is the extent to which the organizational culture in the foreign operation is different from the organizational culture at home. The greater the difference, the more difficult the adjustment and the longer it ultimately takes for the expatriate to adjust.

As anyone who has had to move from one country to another will attest, the logistics involved are tremendous, especially for a family. Thus, the third organizational factor that affects adjustment is the extent to which the organization provides logistical support. By providing this support, a company can significantly reduce the uncertainty associated with finding housing, schooling, and so on. Because much of this support concerns nonwork issues, it tends to have more effect on an expatriate's general adjustment to the nonwork environment than to work.

The fourth organizational factor involves the extent to which members of the foreign operation provide social support to the newcomer. Supportive coworkers can provide both information about how to get along in the organization and emotional support while the newcomer learns the ropes. Recent studies (Feldman & Bolino, 1999) indicate that emotional support may be as critical as logistical support to effective adjustment, especially in collectivist cultures and in those with large power distances. This research suggests that expatriates who receive social support from in-country mentors, for example, are more likely to form positive attitudes toward the work and to complete the assignment than their nonmentored counterparts.

Nonwork Factors. Research has shown that three other nonwork factors contribute to a manager's cross-cultural adjustment. Any of these factors can have a direct influence on the expatriate's performance at work.

The first such nonwork factor is the spouse or partner. Studies show a consistent and strong relationship between how well the manager is adjusting and how well the manager's partner is adjusting (Caligiuri et al., 1999; Harvey, 1998; Pellico & Stroh, 1997). At this point, it is impossible to determine the exact cause-and-effect relation between the employee's and the partner's adjustment. Most likely, the relation is reciprocal so that the adjustment of the employee and the partner influence each other.

Like differences in organizational culture, the greater the general nonwork differences between the native and the host culture, the greater the uncertainty both the employee and the employee's partner have about how to behave appropriately. These differences hinder the employee's interaction and nonwork adjustment more than they hinder his or her adjustment to the job. These differences between the employee's native culture and the host culture determine what various authors have described as the culture's novelty, toughness, or distance.

When there are extensive differences between the employee's native culture and the host culture, we say that there is a great deal of *cultural distance.* Cultural distance has a negative impact on interaction adjustment and on general nonwork adjustment. There are two reasons cultural distance affects adjustment so profoundly. One reason is that the greater the number and degree of differences (distance) between two cultures in such matters as values and ethics, the more potential there is to make cultural mistakes (Lowe, Downes, & Kroeck, 1999). The more depressed and frustrated people become about making these errors, the more defensive and angry they may become toward host-country nationals whom the newcomers often see as the cause of the blunders.

Another reason that cultural distance has a negative impact on adjustment is that the ways in which differences are discovered or learned, the way mistakes are recognized, or how apologies are made may also be different from one culture to another. For example, Japanese and American cultural differences have received a lot of scholarly and popular attention. Business-related differences, such as the use of merit versus seniority pay systems and individual versus group decision making, have been discussed in several popular books. Much of the writing in the academic and popular press suggests that the differences between Americans and Japanese are so numerous and significant that conflict is inevitable. However, conflicts occur among Japanese in Japan and among Americans in the United States. So what is the big deal if conflicts arise between Japanese and Americans working together? The problem is not only that Japanese and Americans do things differently, creating conflicts, but that they also resolve conflicts differently, which in turn can create conflicts that neither side can agree how to solve.

The final nonwork factor that has an important relation to cross-cultural adjustment is the extent to which expatriates receive social support from host-country nationals outside the workplace. Such support can provide both information about how to get along in the culture and emotional support while the newcomers are learning the ropes. As mentioned previously, expatriates with higher levels of support as a result of having a mentor, for example, have been shown to have adjusted better to interacting with the natives and to their general nonwork environment than expatriates who did not receive such support (Harvey et al., 1999).

Influences on Spouses' Adjustment

By comparison with what is known about employees, very little is known about why adapting to life in a foreign culture is easier for some partners than others. What is clear is that unlike the employee, who has the built-in structure of an organization and the routines associated with a job, the spouse—almost always a woman—often has to figure out how to survive and succeed on her own. Not surprisingly, given the feelings of isolation and frustration, depression often sets in and eventually a strong desire to return home.

Pellico and Stroh (1997) conducted a study of spouses in which we tried to systematically examine a number of factors and their impact on cross-cultural adjustment. Pellico and Stroh (1997) found that although managers are often excited about the career potential of an international assignment, their spouses may be a lot less excited. For spouses, a move to a foreign country may simply represent a disruption of their own careers or long-term social relationships. Not surprisingly then, this study found that the more the spouse was in favor of making the move overseas, the more he or she tried to learn about the country and its culture. This self-initiated predeparture or departure training in turn was positively related to the spouse's ability to interact with the locals. Further, those spouses who were interviewed by a human resource professional from their spouses' firm before the assignment had the highest levels of adjustment. This correlation may be explained by the fact that these spouses received a more thorough description of the assignment and country in the interview than they received from their spouses (i.e., the candidates for the assignment).

Given the ego bashing (or at least bruising) expatriates often experience after a cross-cultural move, we expected that whether spouses had social support would be an important factor in their adjustment. Pellico and Stroh (1997) found that receiving support from both the family and from host-country nationals helped the spouses interact better with locals. Support from host-country nationals was particularly helpful, because they could provide information and feedback to the spouse on how well he or she was doing and what changes could lead to functioning more effectively.

Our research identified two other factors that were significantly related to the spouses' general nonwork adjustment. The first of these factors was culture novelty. As with employees, the greater the cultural difference between the home and host cultures, the greater overall difficulty the spouses had in adjusting.

The second factor was a function of the spouses' particular circumstances. Even though many spouses who work full-time before moving overseas find work when they are abroad, most spouses do not work during an international assignment. Consequently, most spouses have to deal with the family's general living conditions all day, day in and day out. It is perhaps not surprising then that having living conditions that are equal to or better than those at home has a positive effect on spouses' general adjustment.

MAXIMIZING THE CHANCES OF SUCCESS

For executives as well as human resource professionals involved in or responsible for the international movement of global managers, a key question is, What steps can the firm take to maximize the chances for success and minimize the chances for failure among our expatriates? This question is especially important because some factors that affect cross-cultural adjustment are not under anyone's control (e.g., the level of organizational or cultural novelty). So what effective steps can be taken? Figure 5.2 summarizes these actions.

Selection

In chapter 3 (this volume), we discussed at length how firms can more effectively select employees for global assignments. Three points are worth reemphasizing in this chapter. First, provide feedback to both candidates and spouses on their cross-cultural strengths and weaknesses through standardized instruments. Knowing their strengths and weakness can help employees and their spouses maximize the effectiveness of any preparation that the employer offers and any they acquire on their own. In addition, the feedback may be an important catalyst and motivator for pursuing educational and training efforts.

Action Category	Action
Selection	• Provide feedback to both the employee and spouse about their cross-cultural strengths and weaknesses through a standardized instrument. • Interview the spouse to investigate the level of motivation he or she has for the assignment. • Take greater care in assessing the employee and the spouse when the assignment is to a highly different culture or to a remote and more harsh locale.
Training	• Provide more rigorous training as the level of communication, job, and cultural toughness increase. • Provide, whenever feasible, both predeparture and postarrival training, and focus the postarrival training on the more complicated aspects of the culture. • Extend whatever training is provided to both the employee and the spouse.
Job Design	• To the extent possible, give the employee considerable discretion and freedom in the job. • To the extent possible, try to clarify the job expectations, responsibilities, and objectives. • Provide overlap with the job incumbent whenever possible. • Know and attempt to align any conflicting expectations concerning the job, objectives, or performance standards.
Logistical Support	• Provide first-class logistical support so that both the employee and the spouse can focus on the challenging and important aspects of effectively living and working abroad. • Monitor logistical programs and solicit expatriate feedback concerning outside vendors.
Social Support	• To the extent possible, encourage local national employees to provide support to new international assignees and families. • Provide comprehensive information about various social, community, or religious organizations that may create opportunities to meet people and develop social support networks.

FIG. 5.2. Effective steps for successful adjustment.

Second, although not all candidates for global assignments are married, most are. Clearly, the spouse's opinion can significantly affect both the decision to accept an assignment and how well the family adjusts after the move. Including spouses and domestic partners in the entire process of relocation—from the selection of candidates through predeparture and postarrival training—is very important in the cultural adjustment of the family and therefore to whether the assignment is a failure or a success. Companies should also make every effort to provide partners with support in the form of mentors, sources of information, and networking opportunities.

Third, although executives cannot control the extent of organizational or cultural novelty expatriates will experience in their host countries, executives can factor it into the process of selecting candidates for expatriation. The greater the cultural novelty, the more careful the selection decision should be.

Training

Although we discussed this topic at length in chapter 4 (this volume), as the level of communication, job, and cultural toughness increase, organizations need to provide more rigorous training. Whenever feasible, this should include training both before departure and after relocation. If training is provided at both stages in the adjustment process, the focus of the predeparture training should be on aspects of daily living in the country, whereas the focus of the postarrival training should be on the more complicated aspects of the culture. If only predeparture training is provided, both daily life and the culture need to be addressed. Finally, as mentioned previously, whatever training is provided should be offered to both the employee and the spouse. Many training providers now have excellent programs for children, and these should be taken advantage of whenever possible.

Job Design

Research shows that positions with greater job discretion and clarity and less role conflict facilitate work adjustment. One might think that it would be relatively easy to ensure job discretion, make role expectations clear, and eliminate conflicting demands when designing overseas positions. However, doing so is actually quite complex.

We consider the issue of role clarity. One easy but rarely utilized technique for increasing role clarity is to build in job overlap, that is, to allow the new expatriate and the incumbent to work together for several days or perhaps a week. During this time, the incumbent transfers necessary information and makes necessary introductions. Americans in countries such as Japan, Korea, Singapore, and Saudi

Arabia in particular have mentioned that time with their predecessors was necessary for making proper introductions to customers, suppliers, business partners, and government officials.

The more complex the job and the less experienced the new entrant, the longer the incumbent and the recently arrived expatriate will need to work together. Theoretically, the incumbent should be able to clarify all aspects of the job. In interviews, several expatriates specifically mentioned that this approach was a relatively low-cost means of facilitating adjustment and effectiveness.

Key executives can also enhance the return on investment (ROI) of an international assignment by reducing role conflict. Such conflict is most likely to occur when the parent firm and the local operation have conflicting expectations. Interestingly, efforts to clarify the job may actually uncover previously hidden role conflict. Thus, firms must consider the need to increase job clarity and decrease role conflict simultaneously. This action requires that both the parent firm and the foreign operation have a clear understanding of what is expected of the expatriate manager and that they work to align those expectations (Au & Fukuda, 2002).

Even a firm's best efforts may not eliminate role conflict and ambiguity. This is probably why discretion tends to be the single strongest factor affecting work adjustment. Even if role ambiguity and role conflict exist, having a fair amount of freedom to decide what tasks to do, how to do them, when to do them, and who should do them facilitates an expatriate manager's ability to cope effectively with ambiguity and conflict. Thus, one way to solve all these problems is simply to give expatriate managers a lot of job discretion and freedom. Unfortunately, this may not be as easy as it sounds. Too much discretion without clear objectives may cause the manager to choose objectives that are not in the best interest of the parent firm, the foreign operation, or both. Therefore, the firm needs to consider all three elements of an expatriate's job simultaneously.

Even though research indicates that job clarity, role conflict, and job discretion are important to work adjustment, we believe that it is best to think of these aspects of the expatriate manager's job as outcomes of broader policy and strategic processes. To make significant, long-term effective changes in the expatriate manager's job, a firm needs to carefully address each of the following issues:

- The objectives of the assignment (Because there are no host-country nationals capable of fulfilling the job? To provide the expatriate with needed developmental experience?).
- The criteria by which success will be measured; in other words, what the firm expects the expatriate to accomplish.
- Whether the objectives and goals of the parent firm and the foreign operation are consistent. The goals of the foreign operation and of the particular department where the individual will work also need to be consistent.

- The amount of control and oversight that will be needed between the parent firm and the foreign operation. Directly related to this issue is the matter of how much freedom and autonomy the foreign unit should be given. Finally, there needs to be consistency between the amount of job discretion the expatriate is being given and the need for coordination between the parent firm and the foreign operation.

Without an assessment of these strategic issues, the firm may adjust its expectations in ways that conflict with the corporation's overall strategy and that are dysfunctional to the relationship between the parent firm and the foreign operation. For example, the incumbent may provide conflicting information about the operation's priorities and objectives, creating expectation conflict, or the firm may give expatriate managers too much freedom when much more control and coordination are required. Adjustments made independently without regard for the broader competitive strategy are likely to have short-term positive results at best and severe negative results at worst. By contrast, an analysis that begins with the broader competitive business strategy and context naturally leads to understanding of and appropriate adjustments to the expatriate manager's job and to a higher probability of successful adjustment at work.

Company Support

Logistical Support. Although most firms would rate themselves as quite good at providing logistical support, employees and families' opinions are not as generous. The critical thing to remember is that excellent logistical support is generally money well spent because it frees the employee and the spouse to focus on other targets of adjustment that yield negative consequences if not accomplished and payoffs if achieved.

Social Support. Evidence consistently indicates that like logistical support, social support can facilitate cross-cultural adjustment for both employees and their spouses. The evidence is particularly strong concerning social support from host-country nationals. Although firms cannot directly control the amount of social support particular families or employees receive, firms can take steps to enhance the probability that expatriates will receive support on their arrival. Some firms have adopted the practice of asking a host-country national to help the newcomers have a "soft landing." Generally, this assistance has focused on day-to-day matters such as housing, schools, and shopping. Although friendships cannot be mandated, some companies have made it clear that helping families (especially spouses) become involved in social or cultural activities is appreciated. Other firms have taken a more indirect approach. They simply provide stacks of information on local social and cultural activities and groups in which managers and their families can become involved.

SUMMARY

In this chapter and chapter 2 (this volume), we discussed the process of cross-cultural adjustment and presented a framework for understanding the complexities of successfully moving human resources around the world. We have emphasized the difficulty and importance of understanding and adjusting to the invisible aspects of culture and the factors that either enhance or detract from that process. We have argued that elements such as organizational and cultural distance or novelty, although not under a firm's direct control, should at least be accounted for in decisions about selection and training. In chapters 6, 7, and 8 (this volume), we focus on expatriates' commitment, performance, and compensation and rewards during global assignments. Each of these components of an international assignment is discussed separately, but these issues are very much interrelated. Those firms that deal effectively with these issues are the most likely to enjoy an advantage in today's highly competitive global market.

CHAPTER

6

Integrating: Balancing Multiple Allegiances

"A house divided against itself cannot stand." Few would argue with the truth of this statement, yet hundreds of global managers have told us that they feel like they are in the middle of a house divided against itself. Literally hundreds of thousands of managers all over the world find themselves in this situation—torn between their allegiance to their parent firm and to their local operation. The cause of all this conflict? In short, these managers are caught between competing pressures, priorities, and processes.

To understand the tension these managers feel, consider the situation. A Dutch manager with a consumer products firm has been relocated to a large developing nation. On one hand, the parent firm wants a particular laundry detergent—in a standard-size box, color scheme, and so on—introduced in the host country as part of a global brand-image strategy. On the other hand, the size and cost of the detergent are prohibitively high for many potential consumers in this developing nation. They are used to scrubbing their clothes with small "bricks" of soap that cost just pennies per brick. The international assignee tries to explain the local situation to management in the corporate office but is branded a nonteam player. When he begins to give in to the global strategy, local managers view him as a corporate crony.

Faced with serving two masters, many global managers end up directing their allegiance too far in one direction or the other, creating serious costs and consequences for both themselves and their organizations. For example, if individuals are too committed to the local operation relative to the parent firm, it is difficult for the home office to coordinate with them. As a senior executive from Honda commented, the carmaker incurred "nontrivial" costs trying to coordinate its global strategy for the Honda Accord because some managers on international assignments were too focused on local situations. Yet managers who are overly committed to the parent firm relative to the local operation often implement policies, procedures, and products from the home office that are inappropriate for the local environment. The medical equipment division of a large U.S. multinational firm recently tried to implement home-office financial reporting and accounting procedures that simply did not apply to and would not work in the company's newly acquired French subsidiary.

Today's multinational firms need managers who are highly committed to both the parent firm and the local operation and who are able to integrate the demands and objectives of both organizations. As one senior executive put it, the bottom-line question is, "How can we get expatriate managers who are committed to the local overseas operation during their international assignments but who stay integrated with the parent firm?"

Unfortunately, very little research has been done on the problems of dual allegiance during international assignments. One of the few studies that has addressed this issue was conducted by two of us and involved more than 300 managers in eight countries (Gregersen & Black, 1992). The good news is that this study provides the initial basis for identifying influential factors, the underlying dynamics, and actions firms can take to more effectively manage some of the challenges international managers face because of their dual allegiance. The bad news is that our research suggests that expatriate managers who have a strong allegiance to both their home and local operations are the exception rather than the rule.

In this chapter, we describe patterns, causes, and consequences of the dual allegiance global managers experience. We start with a look at the four general patterns of commitment or allegiance global managers experience. First, they become overly committed to the parent firm. Second, they become overly committed to the local operation. Third, and ideally, they become committed to both organizations, and fourth, and finally, they become committed to neither the parent firm nor the local operation. These four basic patterns are presented in Fig. 6.1.

Much more important than the patterns of dual allegiance are the factors that cause them and the related organizational and individual consequences. We describe the causes and consequences associated with each cell in the matrix and illustrate them with actual cases generated through numerous interviews and surveys (most managers asked that we not use their names or that of their firms). We also examine what firms are doing now and what they can do in the future to more effectively manage the problems that can occur when managers who are sent abroad develop dual allegiance.

FREE AGENTS

Paul Jackson was a vice president and general manager for the Japan subsidiary of a large West Coast bank. This was his fourth position and the fourth firm he had worked for in the 10 years since he received his master's. Paul majored in Asian

	ALLEGIANCE TO THE PARENT FIRM	
	Low	*High*
Low — ALLEGIANCE TO THE LOCAL FIRM	Expatriates who see themselves as **Free Agents**	Expatriates who leave their **Hearts at Home**
High	Expatriates who **"Go Native"**	Expatriates who see themselves as **Dual Citizens**

FIG. 6.1. Patterns of allegiance.

studies as an undergraduate and spoke and wrote advanced intermediate Chinese when he graduated from college. Paul had also spent 2 years studying in Japan. In that time, his Japanese speaking and listening abilities had reached nearly a professional level. After Paul finished graduate school, he was hired by a major East Coast bank, and 2 years later, he was sent on a 3-year assignment to Hong Kong.

The compensation package Paul and his family received made life in Hong Kong very enjoyable, but Paul felt little loyalty to either the parent firm back home or to the local operation in Hong Kong. First and foremost, Paul was committed to his career. Because he was such a hard charger, the bank invested a substantial amount of time and money in his language and technical training. He worked hard but always kept an ear out for better jobs and pay.

Two years into his Hong Kong assignment, Paul took a better position with another firm. Four years after joining that company, he took a job with a different U.S. bank, working in its operation in Taiwan. Four years later, he took the job in Japan.

In the interview in which Paul related his work history, he said, "I can't really relate to your question about which organization I feel allegiance to. I do my job, and I do it well. I play for whatever team needs me and wants me. I'm like a free agent in baseball or a hired gun in the old West. If the pay and job are good enough, I'm off. You might say, 'Have international expertise, will travel.'"

Hired-Gun Free Agents

We discovered that Paul was part of a network of "hired-gun free agents" in the Pacific Rim. The network consisted of about 10 American managers hired for positions in Asia. To get hired for their current positions, they had to be either bilingual or trilingual and had to have spent more than half their professional careers in the Far East (e.g., Japan, China, Hong Kong, Taiwan, Singapore, Korea, Malaysia, Indonesia). They helped each other by passing along information about various firms that were looking for experienced managers for their Far East operations.

Hired-gun free agents have a low level of commitment to both their parent firms and local operations. They are first and foremost committed to their own "gun-slinging" careers. When asked what long-term career implications this approach might have for them, these managers commonly indicated that it would be very difficult to ever "go back home" and move up the hierarchy in headquarters. They did not consider this a drawback though because most reported that they did not want to go back home for several reasons. First, they felt their children's experiences in their private schools and their general life perspectives were far superior to what they would be back home. Second, the managers said they would be worse off financially if they went home and had to give up the benefits of their expatriate packages. Third, most were confident that they would not be given jobs back home with the status, freedom, and importance of the positions they held overseas. Consequently, most of these hired guns seemed happy with their lives overseas and free-agent careers.

Firms tend to view hired guns with some ambivalence. On the one hand, even though hired guns are given special benefit packages, hiring hired guns often costs the company less than relocating managers from the home country. Furthermore, hired guns have already demonstrated that they can succeed in international settings and have specialized skills (e.g., in the local language) that may be lacking in the firm's internal managerial or executive ranks. This may be especially important to U.S. firms because as discussed in previous chapters, on average, 15% to 20% of managers fail in overseas assignments, primarily because of problems adjusting to the foreign culture. For example, when KFC launched its first restaurant in China, the firm had only three employees who spoke Mandarin Chinese, and only one of them had business experience that even came close to what was required to head up the new venture.

On the other hand, hired gun free agents often quit with little warning. Replacing them is usually costly and difficult, which can have negative consequences for both the parent firm and the local operation. Also, because they tend to be so committed to their own careers, hired guns sometimes take actions that serve their own short-term career objectives but not necessarily the long-term interests of the local operation or the parent firm. Because few hired guns are willing to repatriate, integrating their general international experience or specific country/regional knowledge into the firm's global strategy is next to impossible.

Plateaued-Career Free Agents

We have also uncovered another group of managers who have a low level of commitment to both the parent firm and the local operation. Managers in this group typically have been selected from the ranks of home-country employees rather than of hired international experts. In general, they were not committed to the parent firm before leaving for the overseas assignment, in part because their careers plateaued prior to starting the international assignment. They took the assignment because they saw themselves going nowhere at home and hoped an "international stint" would change things; or they may simply have been attracted by the financial package. Unfortunately, many of the factors that led to low commitment to the parent firm before the international assignment result in low commitment to the local operation once these managers are overseas. We might think of these managers as plateaued free agents.

Several factors can contribute to a manager becoming part of this group. First, if firms allow candidates for overseas assignments to self-select, they open the door for plateaued managers to nominate themselves. As one manager on an international assignment said, "I figured I was stalled in my job [back in North Carolina], so why not take a shot at an overseas assignment, especially given what I'd heard about the high standard of living even mid-level managers enjoyed overseas." As this comment indicates, not all managers who are willing to move abroad have the right motivations or personal characteristics. The chances that the wrong people will nominate themselves increases to the extent that the company does not provide self-assessment feedback through standardized selection instruments. When an unrigorous selection process is combined with a company placing a low value on global operations, the chances increase that plateaued managers will apply for international assignments. At the same time, the chances decrease that high-potential managers will apply (Black, 1989). High-potential managers know that because global operations are devalued, accepting an international assignment is not a way to get ahead.

Lack of predeparture cross-cultural training can also reinforce low commitment to the parent firm and the local operation. U.S. firms may be particularly vulnerable in this regard because, as has been pointed out, roughly 40% to 50% of all American managers who are sent abroad receive no cross-cultural training before they leave; the resulting attitude is, "If the company doesn't care about me, why should I care about the company?" Lack of training can also inhibit managers from understanding the culture and people in the country to which they are relocated and thus from becoming committed to the local operation.

Unlike hired guns, many plateaued-career free agents are not happy in their overseas assignments. Given their low level of commitment, they often make little or no effort to adjust to the local operation and culture. In the most dramatic case, this lack of effort and adjustment can lead to a failed assignment and its associated costs, and not surprisingly, having failed in an overseas assignment is not likely to lead to career advancement, especially for managers who have plateaued. Additionally, failed overseas assignments can strike a severe blow to these individuals' sense of identity and confidence.

As with any failed assignment, there are costs to the firm when a plateaued manager fails. Beyond the $100,000 to bring the employee and family home and send out a replacement, firms incur the costs of damaged client or supplier relationships resulting from poor adjustment and performance while the manager was overseas. Additionally, the "leadership gap" that often occurs during the replacement process can contribute to damaged internal and external relations. Failed overseas assignments may also generate rumors back home that international posts are the "kiss of death" for a career, setting in motion a downward-spiraling vicious cycle.

Even if the plateaued free agent does not return early or otherwise fail, the costs to the person and to the organization of sending a plateaued manager abroad can be high. An interview with a manager of a major U.S. aircraft manufacturer who was stationed in Taiwan illustrates some of these subtle but important personal and organizational costs.

Bob Brown was a typical plateaued manager who was transferred overseas 3 years ago. Bob was not very excited about living in Taiwan and neither was his family. His wife and daughter repeatedly asked to go back home to the West Coast. Bob pointed out that there was really no job for him to go back to. His daughter became so distraught with life in Taiwan that she began doing extremely poorly in school. This and other pressures put a severe strain on Bob's relationship with his wife. In an interview, Bob summed up his situation by stating that his home life was in shambles and that work was merely a paycheck. The parent company and local firm may have been getting their money's worth from Bob because firms in general get their dollar's worth, yen's worth, or pound's worth from these managers, but we doubt it.

EXPATRIATES WHO GO NATIVE

Another pattern of allegiance is found in managers who have high levels of allegiance to the local operation but low levels of allegiance to the parent firm. Because the local operation is embedded in the local culture—its language, business practices, and values—managers in this group usually form a strong identification with and attachment to the larger culture. Consequently, they are often referred to as having "gone native."

Gary Ogden had been with a large computer company for 15 years. Paris was the third city outside the United States where he had been sent. Gary was the country manager for the firm's instrument division in France and had been there for about 18 months. Of his 15 years with the firm, he had spent a little less than half overseas. Given that this was his family's third international stint, it hadn't taken long for Gary, his wife, and their three daughters (ages 6, 9, and 11) to settle in to life in France. Although Gary's French was not perfect, it was decent. His girls, however, were amazing. They had enrolled in regular French schools when they moved to Paris, and now they were fluent for their ages. The Ogdens frequently took trips to museums, to nearby cities and villages, and to other points of interest. In fact, the Ogdens loved France so much that Gary had already requested that his time there be extended even though his contract required him to stay for only another 6 months. When asked to describe his commitment to the parent firm and the local operation, he responded, "My first commitment is to the unit here. In fact, half the time I feel like corporate [the parent firm] is a competitor I must fight rather than a benevolent parent I can look to for support."

Our research suggests that individuals such as Gary Ogden who have spent a number of years overseas and who are skilled at adjusting to local cultures are the most likely to go native. These individuals have a high level of allegiance to the local operation and a relatively low level of allegiance to the parent firm. Part of the explanation for this tendency to go native is that as managers spend more time away from the home office and the home country operations, less and less of their identity is tied to the parent firm (Mowday, Porter, & Steers, 1982). It becomes both literally and psychologically a relatively distant organization compared to the local operation where they now work. This distance, in combination with an ability to relate to and understand the local culture and people, tends to lead to strong identification with the local operation at the expense of the parent firm. Additionally, the lack of formal communication with the home office through mechanisms such as sponsors (individuals assigned to keep in touch with specific managers during their overseas assignments) causes or reinforces high-local and low-parent commitment. Firms that are structured in international divisions and that have cadres of career internationalists may be particularly vulnerable to a high proportion of their managers developing this lopsided commitment pattern. As discussed in this chapter's "From the Front Lines," localization policies become especially critical when a company has a significant percentage of long-term internationalists.

So what are the consequences of going native? Let's consider the case. Gary Ogden often felt that the parent firm was a competitor he had to fight. However, being a smart guy, he did not declare open warfare. He knew that any international assignment was temporary and that his career was to some extent a function of evaluations made about him at the corporate office back home. Consequently, he had to fight the parent firm in subtle ways. "Sometimes I would simply ignore their directives if I didn't think they were appropriate or relevant to our operations. If it's really important, eventually someone from regional or corporate will hassle me and I have to respond. If it isn't important or if they think I implemented what they wanted, they just leave me alone. As long as the general results are good, it doesn't seem like there are big costs to this approach."

On occasion, Gary said, he had had to fight corporate management more overtly. Whereas this may have cost him back home, fighting these fights and especially winning helped him gain the trust and loyalty of the company's French employees. Their greater loyalty made it easier for Gary to be effective in France. His effectiveness often scored points back home with corporate headquarters, which in turn gave him more freedom to do what he felt would lead to good results, including ignoring corporate directives from time to time.

During our interview with Gary, he mentioned that he nearly left the firm shortly after his two previous international assignments were completed. He com-

FROM THE FRONT LINES:

"Localizing International Assignees"
by Jenny Li, Vice President, Human Resources,
International Operations, Sony Pictures Entertainment

In a perfect world, employees are sent on international assignments, train locals to succeed them, and return home, having gained valuable skills that help the company perform more effectively. In reality, however, many expatriates do not return home. Companies must then decide whether, and how, to localize these employees—that is, change their status to that of a local employee.

It is not possible to design a localization policy that will fit all situations. There are too many differences among countries' tax systems, compensation and benefits practices, and labor and immigration laws as well as among employees' personal situations. Companies can, however, take steps to minimize or avoid potential problems by (a) clearly communicating to employees on international assignments that they will be localized after a specified number of years abroad, (b) creating localization guidelines, and (c) providing an alternative for expatriates who do not wish to be localized but for whom other positions are not available within the company.

At the very least, a company should have a clear policy limiting the length of time that an employee may retain expatriate status. In addition, during initial discussions about an international assignment, the company's human resources department and management should communicate the intent to localize after a specified number of years away. Companies may also consider documenting this understanding in employees' contracts.

In developing localization guidelines, companies will need to determine whether and how to buy out or decline allowances, eliminate other expatriate perks, and transition expatriates to local benefits. Retirement, housing, and children's tuition pose more complex problems and typically need to be addressed on a case-by-case basis.

Finally, if an employee does not wish to be localized, companies may want to consider offering the employee a severance package that includes repatriation, outplacement, and separation pay of a sufficient amount to enable the employee to return home and find a position outside the company. Although this is the least desirable outcome following an international assignment, this approach alleviates concerns that potential international assignees may have about accepting international assignments.

plained about the lack of responsibility he experienced back home compared to what he enjoyed overseas. He was also frustrated by the general lack of appreciation and utilization of his international knowledge and experience after repatriation. His low commitment to the parent firm heightened the salience of his justifications for quitting. Gary stated that both times what kept him from quitting was that his request for another overseas assignment was granted.

Negative Aspects of Going Native

From the parent firm's point of view, one of the common problems associated with having managers overseas go native is the difficulty of getting corporate policies or programs implemented at the local level. Because of their intense commitment to the local operation, these managers often implement what they think is relevant, in a way they deem appropriate, and ignore or fight the rest. This can be very costly, especially when the parent firm is trying to closely coordinate activities in a wide variety of countries for the good of global corporate objectives.

Also, to the extent that low commitment to the parent firm contributes to turnover after repatriation, the parent firm loses the opportunity to incorporate these managers' knowledge and experience into the global strategy of the firm or these individuals into the company's plans for succession. Interestingly, our research found that about 90% of managers on global assignments (regardless of their level of commitment to either the local operation or the home office) felt that their firms did not utilize their international knowledge and experience. This may suggest that in general, firms are not utilizing valuable resources in which they have invested substantial sums.

Positive Aspects of Going Native

Despite the various negative aspects of going native, many corporate executives recognize that going native is not all bad. Because of their high level of allegiance to the local operation, managers who go native generally identify with and understand the feelings and values of employees, customers, and suppliers in their host countries and thus get along well with them. This can result in products and services being developed or adapted that are well targeted to the local market. In some cases, this sensitivity to the locals can even lead to ideas and innovations that can be leveraged across the company's global operations. For example, a sensitivity to the need for bigger screens on mobile phones in Japan to accommodate Japanese written characters led Nokia to successfully shift to the use of larger screens on its mobile phones all over the world.

In addition, sensitivity to the local operation often leads managers who have gone native to utilize managerial approaches that are especially well suited for use with host-country employees. The importance of being able to modify one's managerial style in response to the culture of local employees must not be overlooked.

Many companies operate under the assumption that a good manager in Paris will do fine in Tokyo or Hong Kong. Based on this assumption, they select employees for global assignments based primarily on their domestic track records. Unfortunately, the evidence doesn't support this as a best practice. As we have already pointed out in earlier chapters, research shows that managerial characteristics that lead to success at home are not predictive of success abroad (Black & Porter, 1991; Miller, 1973).

Expatriate managers with strong allegiance to the local operation may be beneficial in other ways too. In the case of most multidomestic firms, each overseas unit is likely to compete in its specific national or regional market independent of the firm's other organizational units in other countries. The primary information flow is within the local operation rather than between the local operation and the parent firm or between the local operation and other foreign subsidiaries. At the multidomestic stage of globalization, there is a premium on understanding the local market and the culture of host-country nationals. Managers who go native often have valuable insights into local markets, operations, products, and the effectiveness of various managerial approaches. They often can leverage this knowledge to improve local performance.

In summary, managers who have spent a lot of time away from the parent firm, who are able to adjust to foreign cultures, and who lack formal communication ties back to the parent firm are the most likely to go native. There are both pros and cons to high levels of allegiance to the local operation relative to the parent firm for both the individual and the parent company. Managers who go native often have valuable insights into the local operation, culture, and market and are able to implement or adopt procedures, products, or managerial approaches to fit local situations. However, these managers can also frustrate global coordination efforts and may not be committed enough to the parent firm to stick around long enough after repatriation to pass their knowledge on and have it play a role in country, regional, or global strategic planning.

EXPATRIATES WHO LEAVE
THEIR HEARTS AT HOME

Another group of expatriate managers consists of those with a high level of allegiance to the parent firm but little allegiance to the foreign operation. We refer to these managers as those who "leave their hearts at home." Managers in this group identify much more strongly with the parent firm than with the local operation, the local country, or its culture, language, or business practices.

Earl Markus was the managing director of the European headquarters of the "do-it-yourself" retail division of a large building supplies firm. This was Earl's first international assignment in his 22 years with the company. He was married and had two children in college, neither of whom moved with Earl and his wife when they relocated to Belgium (home of the European headquarters). Earl had worked his way up from a store manager to Southwest regional manager and eventually to vice president of finance over the last 22 years.

The European operations were fairly new, and Earl saw his mission for his 3-year assignment as expanding the number of retail outlets from the current 9 in Belgium to 50 all over Western Europe. The president and CEO of the parent firm, headquartered in the United States, had assigned the chief operating officer, Frank Johnson, to work closely with Earl while he was in Belgium.

One year into the assignment, Earl was on schedule and had opened 15 new outlets in three countries. He was frustrated, however. He had seriously considered packing up and going home more than once during the last year. He claimed Europeans were lazy and slow to respond to directives. When asked about his feelings of allegiance and commitment, he commented that there was no contest. He was first and foremost committed to corporate, and when the next 2 years were up, he was headed back home. As an example of how things had gone, Earl described the implementation of the company's inventory system.

About 8 months into his assignment, Frank Johnson had suggested that Earl implement the new computerized system, which had just been phased into all the American outlets. Frank indicated that he was very excited about the cost-saving and theft-reducing aspects of the new system in the United States and had high expectations of reaping similar benefits in Europe. To operate properly, the system required that sales be recorded daily and that a physical inventory of specific items be conducted randomly every week. These reports needed to be transferred within 48 hr to the central office where total and store-by-store reports and evaluations could be generated. The forms and procedure manuals were printed, and a 2-day seminar was conducted for all the European store managers, the director of operations, and relevant members of his staff. Two months later when Earl inquired about how the system was operating, he discovered it wasn't. He said that all he got from his managers were "lame excuses" about why the system wouldn't work, especially in Belgium.

This case briefly illustrates some of the main causes and consequences of maintaining too strong an allegiance to the parent firm relative to the local operation. Perhaps it is not too surprising that our research found that long tenures working for the parent firm in the home country were significantly related to leaving one's heart at home. All their investment of time, sweat, and heartache had been in the parent firm. This high commitment was in part a function of expecting to receive an ROI from the parent firm in the future. Additionally, over time, the identities of these managers and the parent firm had become intertwined. The natural consequence was a high level of allegiance to the parent company (Glisson & Durrick, 1988; Mowday et al., 1982; O'Reilly & Chapman, 1983).

Our research also revealed that two other factors in combination contribute to this pattern of commitment. First, poor adjustment to the host country and culture, in part fostered by selection processes that consider primarily domestic track records, is important. Because these managers cannot relate to the broader culture and people of the host country, it is difficult for them to feel a strong sense of allegiance to the local operation. Second, having a sponsor back in the home office formally assigned to the manager creates a formal tie to the home office, which in combination with the tie that many years of experience in the parent firm created, directs attention and allegiance toward the parent firm and away from the local operation.

So what personal and organizational consequences result from this pattern of commitment? Earl Markus thought about leaving several times during his 1st year in Belgium, but more than anything his fear of negative career consequences kept him from returning home.

Managers who leave their hearts at home generally fail to identify with the local operation, host country, host national employees, customers, suppliers, and their values. Consequently, they often try to implement and enforce programs that are inappropriate for the local operation, or they implement them in ways that offend, irritate, or frustrate host-country employees, customers, or suppliers. Earl Markus's attempt to implement the inventory system is one such example. His implementation effort antagonized employees and created an adversarial relationship that hampered the implementation of other programs and changes.

Not all the consequences of leaving one's heart at home are negative. Our research found that managers who have a high commitment to the parent firm during international assignments are more likely to want to stay with the firm after they are repatriated. Thus, assuming individuals like Earl Markus gain valuable experience, knowledge, and skills during their assignments, the parent firm has an important opportunity to leverage those benefits after they come home. As a consequence, although the firm might have had only a marginal ROI during the assignment, it can enhance that ROI by leveraging the manager's knowledge and skills after repatriation. Unfortunately, because these managers have a low commitment to the local operation, their knowledge and understanding of the foreign country, its markets, competitors, customers, government, and so on is not as strong as it could be, thus reducing the net return for the company.

Nevertheless, managers who leave their hearts at home often facilitate the coordination of activities between the home office and subsidiaries. For example, because of Earl's "homeward-looking" orientation, it was very easy for the corporate purchasing agent to centralize purchasing activities for the European operations. This coordination resulted in substantial savings that the European operations could not have obtained on its own.

Being able to easily coordinate activities between the home office and subsidiaries can be particularly beneficial for firms at the export stage of globalization. The primary objective of most firms at this stage is to sell products developed and manufactured in the home country in foreign markets, and the primary direction of information flow is from the parent firm to the local operation. Thus, being able to work easily with headquarters may be especially useful for firms at the export stage because of the primacy of home-country operations and the key coordinating role the home office plays. Managers with relatively strong commitment to the parent firm are less likely to resist working with and following the coordination efforts of the home office compared to expatriates with weak commitment to the parent firm.

In summary, there are both pros and cons for the individual and the organization of maintaining a high level of commitment to the parent firm and a low level of commitment to the local operation. Expatriates who leave their hearts at home are relatively easy for people in the home office to work with and can be very valuable in coordinated purchasing, marketing, or other strategic activities. Also, these managers are more likely to stay with the firm after repatriation and can be valuable resources in global strategic planning and succession-related activities. However, their low levels of commitment to the local operation may lead them to make less effort than is usually necessary to adjust to the host country's culture. To the extent that these managers can avoid negative career consequences, they may also come home early, resulting in a host of substantial direct and indirect costs. Finally, these managers may implement home-office policies and procedures that are inappropriate for the local operations or in a manner that offends local employees, clients, or suppliers. These damaged relationships can, in turn, have substantial short-term and long-term negative consequences for both the local operation and the parent firm.

EXPATRIATES WHO SEE THEMSELVES AS DUAL CITIZENS

The final category includes managers who have high levels of allegiance to both the parent firm and the local operation. We describe these managers as dual citizens. The term seems to reflect the active behavior, attitudes, and emotions members of this group exhibit. These managers tend to see themselves as "citizens" of the country to which they have been relocated and of their home country and as citizens of both the local operation and the parent corporation. As dual citizens, they feel a responsibility to serve the interests of both organizations.

Allen Morris was the director of the Japan office of a prominent U.S. consulting firm. This was Allen's second international assignment in his 13 years with the company. His first assignment was a 1-year special project in Singapore 7 years before.

Allen was one of three candidates considered for the job in Japan and was selected based not only on his past performance but on interviews and assessments by outside consultants of his personal characteristics and the demands of the job in Japan. Because the job required working in a very novel culture and a high degree of interaction with host nationals, Allen was given 5 months notice before he had to move. During this time, he received about 60 hr of cross-cultural training, and his wife received about 10 hr of survival briefing. About 4 months after relocating, Allen received another 40 hr of cross-cultural training specifically related to Japan—its culture, business practices, and so forth. He also took advantage of more than 250 hr of language training, paid for by the parent firm, after he arrived in Japan.

Allen had a clear set of objectives for his global assignment. The Japan office of the consulting firm had initially been established to serve the Japanese subsidiaries of their U.S. clients. Growth of the Japan office was limited by the slowed pace of expansion of U.S. client firms to Japan. Allen was charged with developing Japanese clients. This would serve two objectives. First, it would increase the growth potential of the Japan office, and second, it would make it easier to secure U.S. subsidiaries of Japanese firms as clients and facilitate the growth of the company's large operations in the United States.

In addition to a clear set of objectives, Allen found there was relatively little conflict between the expectations placed on him by the parent firm and by the local operation. Significantly, people in headquarters realized the time and expense needed to cultivate effective contacts and relationships in Japan. Unlike some of his friends in other firms, Allen did not feel any tension between the firm's "bean counters" who went crazy over entertainment expenses in Japan and local staff who constantly floated "contact opportunities" by Allen.

In addition, Allen understood how his assignment fit into his overall career path and how his repatriation would be handled. Although he was not guaranteed a specific job or position when he returned to the United States, Allen knew what the process of repatriation would involve and what general opportunities would be his if he met his objectives for his international assignment.

Perhaps most important, Allen was given a great degree of discretion and autonomy in how to achieve his objectives. According to Allen, this often gave him the flexibility to deal with the inevitable conflicts between the parent firm and the local operation or the various ambiguities that cropped up in his job.

When asked about his commitment to the parent firm and to the local operation, Allen commented, "I feel a strong sense of allegiance to both companies [local operation and parent firm]. Although they sometimes have different objectives, I try to satisfy both whenever I can." When objectives or expectations were in conflict, Allen would generally work to integrate the needs of both organizations rather than simply following one or the other.

The personal and organizational consequences of Allen's dual-citizenship orientation were primarily positive. At a personal level, Allen indicated that it was sometimes frustrating to feel torn in two directions because the needs of the parent firm and the local operation conflicted. He added, however, that the clarity of his objectives, the latitude he had to pursue them, and the relative infrequency and small magnitude of the conflicts made working to benefit both organizations rewarding and personally satisfying. Allen did well in his 5-year assignment in Japan and received a promotion on repatriation to a position in which some of what he had learned was utilized. At the organizational level, Allen's dual-citizen orientation facilitated the development of solid relationships with Japanese clients as well as with government officials and facilitated efforts by the home office to establish relationships with the U.S. subsidiaries of new Japanese clients. Allen felt his dual focus enabled him to achieve other results as well, including greater ability to recruit high-quality Japanese employees (something competitors struggled with).

In our research, we have found that roughly one fourth of managers on international assignments have high levels of commitment to both the local operation and parent firm. Although it would be inaccurate to say that these managers never returned home early, never left the firm after repatriation, or never had adjustment or performance problems during their international assignments, their dual allegiance led to a higher probability that these managers would stay in their assignments for the expected length of time, stay with the firm on repatriation, and adjust well during their stays overseas. These managers were sincerely interested in understanding the needs, objectives, constraints, and opportunities of both the local operation and the parent firm. They talked of trying to utilize this understanding to find solutions that would satisfy and benefit both organizations. This approach created the possibility of not only effectively implementing policies in the local operation that were created by corporate officers but also of passing along information and guidance to corporate headquarters that could help in shaping more effective strategy and policy.

Although dual-citizen managers are desirable for any firm at any stage of globalization, they are most critical for firms at the coordinated multinational stage. Firms at this stage need information to flow easily between the home office and their foreign subsidiaries and from one foreign subsidiary to another. They need managers who identify both with the people at the parent firm and those in the local operation. They need managers who try to integrate and meet the needs of both organizations. They need managers who are going to complete their international assignments. Finally, they need managers who will stay with the firm after repatriation so that their international experience, knowledge, and skills can serve as assets in the global strategy and policy-making activities of the corporate office. Managers who are highly committed to both the parent firm and the local operation are critical to fulfilling these needs.

Figure 6.2 summarizes the positive and negative aspects of each of the four patterns of commitment.

GUIDELINES FOR THE EFFECTIVE MANAGEMENT OF DUAL ALLEGIANCE

Although most multinationals and their executives are aware of the issues concerning dual allegiance among international managers, very few executives we have interviewed thought their firms had a clear understanding of the causes and consequences of the different patterns of commitment or a systematic means of developing managers who had dual commitment. Instead, many firms seem to have found ways to counterbalance tendencies to become overly committed to one organization or the other. In addition to presenting what some firms are doing to counterbalance "lopsided" allegiance, we also propose, based on our research

Allegiance Pattern	Pros	Cons
Free Agent	• Typically have superior and demonstrated international capabilities (e.g., language, negotiation, management). • Often somewhat less costly than traditional expatriates.	• Often leave with little warning. • Replacement costs may be significant. • May serve self-interests more than company interests.
Go Native	• Typically adjusts well and quickly to the local culture and environment. • Usually is effective in the local environment, including interactions with employees, customers, suppliers, and government officials.	• May fight global initiatives. • May be slow to implement directives from headquarters. • More likely to leave the firm after repatriation.
Heart at Home	• Facilitates the coordination of global initiatives. • Quick to implement directives from headquarters. • More likely to stay with the firm after repatriation.	• Typically adjusts poorly and slowly to the local culture and environment. • Often is not effective in the local environment managing employees, customers, suppliers, or government officials. • Likely to inappropriately implement directives from the parent organization.

FIG. 6.2. *(Continued)*

Allegiance Pattern	Pros	Cons
Dual Citizen	• Typically adjusts well and quickly to the local culture and environment. • Usually is effective in the local environment, including interactions with employees, customers, suppliers, and government officials. • Facilitates the coordination of global initiatives. • Responsive to directives from headquarters. • More likely to stay with the firm after repatriation.	• They require serious thought and commitment from the company to create. • They are a rare breed and may be quite attractive to other firms who may try to steal them away.

FIG. 6.2. Allegiance pattern pros and cons.

results, steps firms can take to develop managers with balanced, high levels of allegiance to both the local operation and the parent firm.

Strategy 1: Matching Subsidiaries and Expatriates

The thrust of this strategy is based on the reality that not all companies are at the same stage of globalization. As a consequence, the strategic role of foreign subsidiaries varies from company to company. This variation means that a "one-size" manager does not fit all global situations. In our research and work with companies, we have found that a few recognize this and thoughtfully match candidates for global assignments based, at least in part, on the goals of the subsidiary.

You may recall that in chapter 1 (this volume), we introduced the idea that a given foreign subsidiary may have a particular strategic function relative to the flow of information. To remind you, the diagram that we presented in chapter 1 (Fig. 1.1, this volume) is reproduced as Fig. 6.3. Island subsidiaries focus on the local market. Information, technology, and knowledge do not flow out of the home office or into the local operation in heavy volume. Implementor subsidiaries

Information Flow

	Low Flow In	High Flow In
Low Flow Out	Island	Implementor
High Flow Out	Innovator	Integrator

FIG. 6.3. Foreign subsidiary strategic function.

take information, technology, knowledge, and products from outside (typically from the home office) and implement them in the local market. Innovator subsidiaries serve the local market but focus on the world market. Innovators generate information, technology, knowledge, and products and send those innovations to the parent organization and to other foreign subsidiaries. Integrator subsidiaries focus on both the local and world markets. Information, technology, and knowledge flow both into and out of them in a high volume.

Given these different strategic functions and different pros and cons of the four commitment patterns described earlier, we now examine how managers with different commitment patterns match up to the needs of subsidiaries at different stages of globalization. Figure 6.4 illustrates this match-up.

Because island subsidiaries are focused on the local market and both inflows and outflows of knowledge are low, these subsidiaries benefit most from manag-

FIG. 6.4. Strategy and allegiance alignment.

ers who have skills and connections specific to the local operation such as language skills, cultural knowledge, political contacts, and so on. In this context, the negative aspects of free agents are not a problem. However, their positive attributes, such as their language skills, experience, cultural knowledge, and so on, are well matched to meeting the needs of island subsidiaries. Although it is understandable that some firms choose not to hire free agents for any positions, others recognize that free agents are well suited for management positions in island subsidiaries.

Implementor subsidiaries need managers who have connections with and who are orientated toward the source of the information, technology, knowledge, and products that are being implemented in the local market. Although heart-at-home managers would likely be poor fits in island subsidiaries, they are often good fits for implementors. Their homeward orientation makes it possible for heart-at-home managers to connect and work effectively with the source of what these subsidiaries are bringing in and implementing.

Innovator subsidiaries draw from the local culture to produce information, knowledge, products, and innovations that are sent to the rest of the organization. This may be because of a high concentration of resources in that locale or what economists like to call "factor endowments." For example, a large Japanese chemical firm's subsidiary in Southern California was expected to draw on the pool of high-quality scientists produced at two nearby universities. Whereas heart-at-home managers would be poor fits in innovators, managers who have gone native would be good fits. Their natural affinity for and knowledge of the local environment can be just what innovator subsidiaries need to tap into local talent.

Integrator subsidiaries have heavy flows of information, technology, knowledge, and the like both in and out and therefore have highly interdependent relationships with the home office and with other units of the organization around the world. Free agent managers, heart-at-home expatriates, and managers who have gone native do not match the needs of integrator subsidiaries well. Dual-citizen managers are a good match, though. Although dual-citizen managers might work well in any of the four subsidiaries we have described, no other type of manager would work as effectively in an integrator.

This strategic alignment or match-making perspective leads to several important implications. For example, not all firms need the same percentage of dual-citizen managers. Firms that have a primarily multidomestic orientation need a smaller percentage compared with firms that are dominated by an integrated, global posture. In our experience, firms in the latter situation tend to have too few dual citizens, or what others call global managers. Another important implication is that because dual-citizen managers are the exception rather than the rule and cannot be developed overnight, most firms find that they need to spend some time identifying and developing them. Again, in talking with literally hundreds of executives around the world, we have found that most find that companies that make this effort are the rare exception and not the rule.

Strategy 2: Counterbalancing Going Native

Even if a firm does the best it can to match managers with assignments, it still may find that too many of its managers have gone native. In this case, the immediate question is, "How can we effectively counterbalance this tendency?"

Managers who are most likely to go native are those who have several years of international experience. They also tend to be successful at adjusting to the general nonwork environment of other cultures. Ironically, these managers are also good candidates to send overseas, in that they have demonstrated that they are survivors (Black, 1988).

One action Honda Motors takes to try to counterbalance tendencies to go native is to have managers return home to Japan for a few years before they are sent overseas again. This practice reinforces the link between the individual manager and the parent firm and counteracts the tendency to become overcommitted to the foreign operation. Honda's view is that it is not logical to expect career internationalists who move from one global assignment to another to be highly committed to the parent firm unless the parent firm deliberately does something to strengthen the tie between the firm and the manager.

Firms can also counterbalance the tendency to go native by selecting managers for global assignments who have longer rather than shorter tenures in the parent firm. As mentioned, the longer managers have been with the parent company, the more they have invested in it and the more they identify themselves with it, the more they are committed to it. However, this recommendation is problematic for firms such as GE, General Motors (GM), and Ford, which increasingly utilize international assignments as developmental experiences for younger, high-potential managers. Consequently, GE takes a broader approach to counterbalancing tendencies to go native. In some GE divisions, this even involves making a commitment to hire the manager back to a specific position before he or she even leaves for the global assignment. More often, GE's program involves the following six elements:

1. Assessing the expatriate's career objectives.
2. Choosing a senior manager (often in the function to which the expatriate is likely to return) who is willing to serve as sponsor.
3. Maintaining contact between the sponsor and the manager throughout the assignment including face-to-face meetings.
4. Clarifying the manager's career objectives and capabilities before repatriation.

5. Evaluating the manager's performance during the assignment.
6. Providing career advice as well as helping the manager find a position before he or she is repatriated.

We talked with executives at several firms with sponsorship programs. In addition to the preceding advice, they recommended that sponsors should be systematically assigned. The sponsor should be senior enough to the manager going on the international assignment that the sponsor can provide a broad view of the organization. To aid in this process, the sponsor should be given specific guidelines about keeping in touch with the manager including such matters as the form, content, and frequency of contacts. Too often the sponsor is simply assigned, and that is it. If the sponsor takes the initiative and fulfills his or her responsibilities, things go well. Otherwise, the sponsor is that in name only. Finally, the responsibility of planning for the manager's return and of finding a suitable position should not rest solely on the sponsor but must be incorporated into the career systems of the firm.

Although most U.S. firms do not provide cross-cultural training before or after managers are relocated, our research indicates that cross-cultural training is an effective mechanism for counterbalancing tendencies to go native. Although it might seem that predeparture training would help managers identify more closely with the host culture and thus increase the probability of going native, evidence consistently suggests that in fact, good predeparture training helps managers adjust to their jobs overseas and increases their performance. Our data indicate that training has an even stronger impact on the sense of obligation and commitment managers feel toward their parent companies. This is because even more than culture-specific content, training demonstrates to managers going abroad that the firm cares and is concerned about them. Managers reciprocate with stronger allegiance to their firms. Although our research examined only the impact of predeparture training on managers' allegiance (because less than 10% of managers receive postarrival training), we suspect that the results would have been similar if training had been provided after arrival, especially if the training were clearly seen as being paid for and sponsored by the parent firm and not the local operation.

In summary, firms can counterbalance the probability that managers will go native in the following four ways:

1. Have managers returning from international assignments work in the home office for several years before sending them on another global assignment.
2. Whenever possible, select managers for international assignments who have a relatively long tenure with the firm.
3. Institute systematic sponsorship programs for managers going on international assignments.
4. Provide predeparture or postarrival cross-cultural training for managers.

Once companies have undertaken these shifts in policy, they should see an increase among their managers in their levels of commitment to both the local operation and the parent firm.

Strategy 3: Counterbalancing the Tendency to Leave Your Heart at Home

Executives in U.S. firms seem less concerned with counterbalancing the tendency for managers to leave their hearts at home than in counterbalancing the tendency to go native. In both cases, however, the negative consequences can be serious. Managers who are most likely to leave their hearts at home typically have many years of tenure with the parent firm and little international experience. Thus, firms such as GE, GM, and Ford, which, as previously mentioned, are increasingly sending younger managers overseas as part of their career development, are perhaps unintentionally counterbalancing the tendency for managers on global assignment to leave their hearts at home.

Helping managers on international assignments adjust to their general nonwork environments is another powerful way to counterbalance the tendency for managers to leave their hearts at home. Ironically, many of the perks (a company car and driver, company housing, etc.) that senior executives on international assignments are given may serve to isolate them and inhibit their adjustment overall. Also, as discussed in chapter 5 (this volume) especially, a manager's adjustment is closely linked with the family's adjustment (Black, 1988; Shaffer et al., 1999). Family members (especially spouses or domestic partners) are often more directly exposed to the general environment because they do not enjoy the insulation that the corporate structure provides for the employee. Thus, company efforts to facilitate the family's adjustment can have a positive effect on the manager's adjustment to the general environment and counterbalance the tendency to leave his heart at home.

The question then arises, "How can firms facilitate the family's adjustment?" One factor that helps families adjust helps managers adjust as well—that is, interacting with host-country locals. The more families interact with locals and, in particular, the more managers interact with them outside work, the more likely these managers will be to adjust to living and working in the country. Providing on-site mentoring facilitates managers' adjustment and heightens their commitment to complete the assignment (Feldman & Bolino, 1999). It doesn't take a rocket scientist to figure out why this is so. Locals know how to get along in their own culture. Consequently, they are the best sources of instruction and especially feedback on getting along on a daily basis.

Ford Motor is one of the few American firms that tries to consistently provide training and preparation for the families (especially spouses) of managers going on international assignments. Although the goal of this effort was not to counterbalance leaving your heart at home tendencies, our research suggests that this is a likely consequence of the company's training and preparation program.

A preparation program can facilitate interaction between host-country nationals and newly arrived managers and their families, but it cannot guarantee that interaction will take place. An additional step firms can take is to ask host-country employees and their families to help out specific expatriates and their families during the first few months after their arrival. Care should be taken to match host-country families and relocated families on such characteristics as the number of children in the family, their ages, and so forth. In one rather extreme case, several Japanese auto firms hired Americans who spoke Japanese to help newly relocated managers and their families adjust to life in America.

Strategy 4: Creating Dual Citizens

There are several steps firms can take to develop managers with dual allegiance. The work environment represents the single most powerful element in doing this. The specific steps to be taken may seem trivial or obvious at first; however, they are likely to require most firms to expend considerable effort in that the focus is on enhancing role clarity and job discretion and reducing conflict.

Enhancing job clarity, reducing role conflict, and determining the appropriate level of discretion are not as simple as they may seem. However, nothing is more effective at fostering dual citizenship. Further, it is important to keep in mind that you cannot turn these "dials" independently; changes in all three elements of the international position must be made simultaneously.

The assessment and design of international positions must take into account not just its technical aspects but the strategic dimensions as well. Without an assessment of the strategic dimensions, firms may unintentionally design in dysfunctional job requirements. In contrast, an analysis that begins with the broader strategy and context naturally leads to understanding and appropriate design of the job and a higher probability of a dual allegiance.

Some readers may be saying to themselves, "We are a global firm, and we need managers who are not just capable of dual citizenship but of world citizenship. We need global managers." Clearly, many firms are moving in this direction. Our data suggest that despite the glamour of world citizenship, most managers who live abroad struggle to reach dual allegiance.

The first practical step in creating global firms and global managers is to develop managers who at least see themselves as dual citizens. This may be especially critical for firms that have reached or are working to reach the coordinated multinational stage of globalization. In this case, developing dual citizens is most likely to be accomplished by taking four specific actions:

1. Selecting candidates based on their cross-cultural abilities, not just their technical capabilities.

2. Providing prearrival and postarrival cross-cultural training.
3. Instituting career systems with clear, consistent job expectations and appropriate levels of freedom and discretion.
4. Providing repatriation programs that reintegrate returning managers and effectively utilize their knowledge, skill, and experience.

These steps will help firms more effectively manage their managers with dual allegiance and help managers on global assignments more successfully serve two masters.

CHAPTER
7

Appraising: Determining if People Are Doing the Right Things

Wayne Casio (1986), a longtime researcher in human resource management, best sums up the problematic nature of performance appraisal systems when he says that a performance appraisal is a multifaceted process. Sometimes it is used to observe and judge employee behavior, sometimes it is used as a development tool, and sometimes it is used as an organizational intervention. In each of these cases, the appraisal involves measurement of behavior that is highly susceptible to inexactness due to human error (Casio, 1986). In this chapter, we address each of these facets of performance appraisal. We then examine them from a global perspective to determine what unique challenges arise in the development of performance appraisal systems for global managers. Finally, we offer prescriptions for companies to assist them in constructing sound performance appraisal programs for managers who are overseas.

PURPOSE OF PERFORMANCE APPRAISALS

Organizations conduct performance appraisals for two main reasons: evaluation and development. Unfortunately, these two purposes are often mutually exclusive and cause friction within the organization.

There are three main evaluation goals for performance appraisal systems (Beer, 1981; Casio, 1986):

1. To provide feedback to managers so they will know where they stand.
2. To develop valid data for pay and promotion decisions and to provide a means of communicating these decisions.

ARE PEOPLE DOING THE RIGHT THINGS? 145

3. To help management in making discharge and retention decisions and to provide a means of warning subordinates about unsatisfactory performance.

Now, consider the development goals of performance appraisal systems (Casio, 1986):

1. To help managers improve their performance and develop their future potential.
2. To develop commitment to the company through discussion of career opportunities and career planning.
3. To motivate managers via recognition of their efforts.
4. To diagnose individual and organizational problems.

As Fig. 7.1 illustrates, these two sets of goals often come into conflict. When a performance appraisal is being conducted for evaluation purposes, the appraisal is a tool by which managers make difficult judgments that affect their subordinates' futures. This function can create an adversarial relationship and low trust, which then works against the coaching and development purposes of the appraisal (Beer, 1981).

Figure 7.1 also illustrates the conflict that can occur in employees who are being evaluated. They desire feedback about their performance to reduce uncertainty about their standing in the organization; yet conversely, they also want to hear only good news. This psychological tug-of-war can exacerbate poor performance, as worry and stress set in and the focus seems to shift to appearance versus substance and quantity versus quality of performance. These stresses and strains, caused by the competing goals for the appraisal in the organization and the resul-

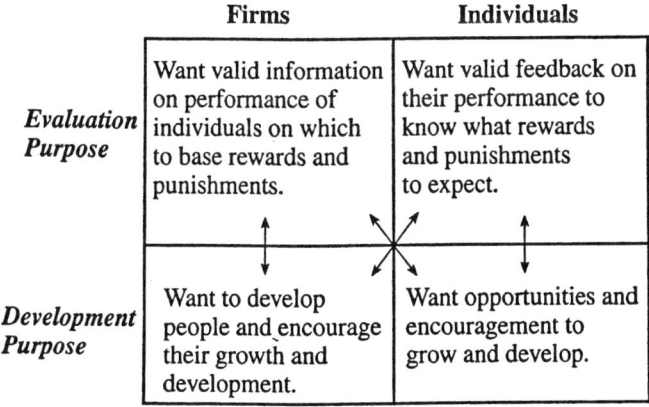

FIG. 7.1. Conflicts in performance appraisal.

tant competing needs within individual employees, are a partial cause and reinforce the knotty dilemmas and problems encountered when designing and implementing performance appraisal systems.

CHALLENGES TO THE DESIGN OF VALID APPRAISAL SYSTEMS

We would need an entire book to cover all the issues of performance appraisals, so this section focuses solely on "problem dimensions." Within each of these dimensions are a large number of specific issues that cause performance appraisal systems to fail. These dimensions are really "umbrella" concepts that encompass a variety of dysfunctional organizational behaviors. Seasoned managers know that these dimensions are virtually impossible to totally eradicate in organizations. Often, the best an organization can do is minimize the extent to which these dimensions hurt organizational and individual growth and performance.

Invalid Performance Criteria

It is not unusual for companies to measure people on behavior that actually does not assist the organization in attaining its goals. For example, it is probably not logical to give significant weight to punctuality in arriving at work as a performance criterion to measure the job performance of a software designer or a commercial artist, yet some companies include this dimension on standardized appraisal forms.

Isolating the key factors of success for any managerial position in a company is not as easy as it seems. The more complex the job, the more difficult it is to isolate the behaviors that enable someone to perform in an outstanding manner in a given position.

Rater Competence

The rater must have the technical background and the experience necessary to evaluate employees' behavior correctly. When raters have limited contact with those whom they evaluate, invalid ratings are likely to occur. The same holds true if raters do not have a clear idea of the actual complexities of the work environment. For example, it would be foolhardy to expect a manager who only understood sales results to effectively evaluate salespeople who must develop junior salespeople, jointly design systems with clients, and work with marketing research. A complex system of "little things" often must be done well for a task to be accomplished; thus, lack of understanding of the complexity of the job tends to cause inaccurate appraisals.

Rater Bias

Rater bias occurs when the rater's values, perceptions, and prejudices replace organizational standards as the basis for evaluation. This bias can take many forms. For example, a manager may "protect" a valued subordinate by giving him or her lower ratings than deserved so that the subordinate will not be promoted or transferred from the rater's organizational unit. A rater may also give an employee lower rating than deserved because of how the employee dresses, speaks, or generally acts around the office. Thus, personal preference "trumps" performance standards.

Because raters are to some degree evaluated on the performance of their subordinates, it is in the raters' best interests to have subordinates who are "above average." As a result, entire staffs may receive undeserved positive ratings. Another rater bias is reflected when recent events (positive or negative) get more emphasis in evaluations than past events do; the longer the interval between evaluations, the greater the effect of this kind of "recency bias." In short, rater bias can take many forms depending on the personal needs of the rater.

CHALLENGES IN THE INTERNATIONAL CONTEXT

The problem dimensions just outlined exist in the international arena as well as in the domestic context. The pertinent issue, however, is whether the manifestations of these problems are different or more severe in the international context. To answer this question, we need to explore the international aspects of each problem dimension. With the assistance of nearly 70 multinational firms, two of the authors of this book (H. B. Gregersen and J. S. Black) conducted one of the few studies that has evaluated performance appraisals of managers on international assignments (Gregersen, Black, & Hite, 1995; Gregersen, Hite, & Black, 1996). The findings reported in the next several sections of this chapter are based on that study as well as on other recent studies that have investigated the effectiveness of performance appraisals of managers working abroad (Bonache, Brewster, & Suutari, 2001; Caligiuri, 2000; Caligiuri & Day, 2000; Lindholm, Tahvanainen, & Bjorkman, 1999; Selmer, 1997; Suutari & Brewster, 2001; Tahvanainen, 1998; Woods, 2003). The "From the Front Lines" in this chapter describes some changes AT&T made as part of a strategy to improve professional development among its international employees.

Invalid Performance Criteria

For a performance appraisal system to work, a clear link must be established between what it takes to be successful on the job and what the appraiser is measuring. Too often, the criteria for success at home are not relevant in another setting (Bonache et al., 2001).

The common performance criteria on which managers are measured at home involve profit and loss, rate of ROI, cash flows, efficiency (input–output ratios),

> ### FROM THE FRONT LINES:
>
> ### "Performance Development for International Employees"
> by Richard R. Bahner, R. Bahner International
>
> When I first began managing AT&T's international business, performance development for employees on global assignments was seriously lacking. In response to this problem, I instituted a multilevel approach. As a first step, we made it a requirement that each long-term international assignee had to complete an Individual Performance Development Plan with his or her supervisor before going abroad. If the assignment changed or the employee completed 2 years of the assignment, the plan was reviewed and adjusted by line management with human resources support.
>
> Each employee who worked abroad was also required to have a mentor with whom he or she would consult. This change ensured that performance development would start taking place.
>
> AT&T also made changes in its compensation package for global employees. As an example of one important change, employees were given only 30% of their international service premium and other incentives before they relocated. They did not receive the other 70% until 6 months after they returned home and after they had completed and presented the Assignment Summary Document. If at the end of the 6 months, the individual was not in an assignment that leveraged his or her strengths and experience, I could reassign the employee to a unit that would benefit from the employee's skills. The record of the employee's accomplishments during the assignment and the Assignment Summary Document were extremely useful in pitching the employee's potential value to another unit.
>
> By delaying payment of international service premiums, AT&T provided a strong incentive for employees to stay with the organization for at least 6 months after returning from global assignments. This new policy also saved AT&T considerable money because incentive payments were free from foreign taxes.
>
> In combination, these various changes in policy increased the attention given to lowering the attrition rate among AT&T's repatriates. The changes also ensured that professional development would be a key component of any stint abroad.

market share, conformity to authority, and physical volumes. It would seem on the surface that managers on international assignments should be evaluated on these same criteria because they are easily measured and straightforward enough to lead to clear performance standards. After all, companies want their managers to achieve the same goals no matter where they are in the world. As Brewster (1991) claimed, this is indeed how a majority of firms measure their global managers' performance. However, simply inserting the same criteria from a performance appraisal at home in one abroad can cause problems.

One reason it is unfair to evaluate global managers against domestic performance standards is that external factors often influence the financial and organizational performance of the international assignee's area of responsibility. The following environmental factors are just some that can distort the appearance of financial and other performance standards in ways that staff in the home office often cannot foresee:

- Rapid exchange-rate fluctuation.
- Price controls.
- Control over the revaluation of assets.
- Depreciation allowances.
- Costs assessed by the parent company and other associated firms against the foreign subsidiary (e.g., price of materials, general overhead charges).
- Availability of local debt financing.
- Local currency evaluation of foreign-source assets invested.

The problems of assessing current profit and loss from liquidation of assets and investments can sometimes cause home-office evaluators to make invalid performance evaluations of managers who are working abroad (Robinson, 1978). For example, severe inflation for months or years with no devaluation of the local currency can help a subsidiary earn high profits, but the profits are more attributable to the inflation rate than to good management. Conversely, when currency is devalued against the home currency, the subsidiary, although well managed and profitable in terms of local currency, may show a loss in a given accounting period when its income statement is translated into the home currency. The experience of a global manager who was stationed in Chile illustrates this problem:

> In Chile he had almost single-handedly stopped a strike that would have shut down their factory completely for months.... In a land where strikes are commonplace, such an accomplishment was quite a coup, especially for an American.... However, because of exchange-rate fluctuations with its primary trading partners in South America, the demand for their ore temporarily decreased by 30% during the expatriate manager's tenure. Rather than applauding the efforts this expatriate executive made to avert a strike and recognizing the superb negotiation skills he demonstrated, the home office saw the expatriate as being only somewhat better than a mediocre performer. (Oddou & Mendenhall, 2000, p. 215)

This global manager's home office placed important emphasis on sales figures without understanding the context that influenced sales in Chile. As a result, all of the manager's other accomplishments were downgraded. Unfortunately, only 10% of the firms surveyed actively incorporate contextual factors in their appraisals (Gregersen et al., 1996).

A foreign subsidiary may be measured on traditional criteria, but corporate headquarters may make it difficult to achieve the performance standards for several reasons such as delays in decisions from headquarters, cumbersome reporting procedures that headquarters has superimposed on the subsidiary, or complete disregard of suggestions made by the subsidiary for changes that would enhance its success. Transfer pricing can force some subsidiaries to show profits that have been allocated to them but that they have not really earned. Global managers of the subsidiaries from which the profits were actually taken may be evaluated less highly than they deserve, especially if staff members at regional headquarters are doing the evaluating and have not been informed by corporate headquarters about transfers that have taken place.

The evaluation goals of performance appraisals are to isolate the truly great performers in a company and to reward them for their efforts. When this does not happen, word gets around, and instead of focusing on what really needs to be done, managers on global assignments begin to play games with statistics to make themselves look good to the home office. Sometimes the real keys to success in international settings are unique, such as the following:

- Fostering personal relationships with local government leaders.
- Fostering personal relationships with union leaders.
- Focusing on the public image of the firm in the local environment.
- Focusing on local market share versus ROI or profit targets.
- Paying constant attention to employee morale and job satisfaction.
- Possessing excellent interpersonal negotiation and cross-cultural skills.
- Spending time and money on community-involvement initiatives.

Many of these important aspects of a global manager's job are not easily quantified or measured, but if they are not measured and evaluated, managers soon learn that it does not pay to concentrate on them.

Armstrong (1994) contended that there should be balance in a performance appraisal in the measurement of "achievements in relation to objectives, behavior on the job as it relates to performance (competencies), and day-to-day effectiveness" (p. 93). To date, research indicates that the criteria used to measure the performance of managers on global assignments are not balanced, and criteria that capture their unique work challenges are often not used (Brewster, 1991; Gregersen et al., 1996; Lindholm et al., 1999; Woods, 2003).

Rater Competence

In domestic settings, managers are evaluated by people who work fairly closely with them and with whom they have had a fair amount of interpersonal interaction. In global settings, the situation is usually different. The global manager may be evaluated by a regional, area, or corporate executive with whom the manager has little face-to-face contact. Also, the evaluator may have little international experience and therefore may not understand the totality of the manager's work situation. Gregersen et al. (1995) found that only 25% of the top executives had international experience and only 11% of human resource managers had such experience. In 21% of the firms, none of the top executives had international experience; and in 59%, none of the top human resource executives had international experience. Yet these same executives may feel perfectly capable of evaluating international assignees.

The previously cited example of the manager in Chile illustrates this problem. He was evaluated by someone in the home office who did not understand Chile's business environment. A high-potential manager sent to Japan by a large semiconductor firm had the following experience, which illustrates the dangers of having a poor rater evaluate a global manager (Oddou & Mendenhall, 2000):

> He barely kept his head above water because of the difficulties of cracking a nearly impossible market. On returning to the United States, he was physically and mentally exhausted from the battle. He sought a much less challenging position and got it because top management then believed they had overestimated his potential. In fact, top management never did understand what the expatriate was up against in the foreign market. (pp. 215–216)

Global managers frequently indicate that their home offices deal with them from an out-of-sight, out-of-mind philosophy and do not understand what these managers encounter overseas. In a 1981 Korn-Ferry International report, 69% of the global managers surveyed reported that they felt isolated from their domestic operations and home-office superiors. More recent research has found similar results (Gregersen et al., 1995).

A competent evaluator who understands the reality of the business situation can compensate for a poorly designed performance appraisal system that measures the wrong things, but in combination, a poorly designed system and a rater who does not understand the business situation of the ratee almost inevitably leads to invalid performance evaluations. When top management does not understand the realities of the overseas business environment, invalid performance evaluations, either formal or informal, are liable to occur.

Rater Bias

People from different cultures often judge one another's behavior quite differently (Selmer, 1997; Stening, Everett, & Longton, 1981). For example, a research study (Stening et al., 1981) that investigated how global managers from Britain, Japan, and the United States and their host-country employees in Singapore perceived one another found the following:

1. The American and Japanese managers perceived themselves to be more technically competent than the British and Singaporeans perceived the Americans and Japanese to be.
2. The British managers perceived themselves to be more technically competent than the managers in the other groups perceived them to be.
3. The American managers saw themselves as being more open interpersonally than the British or the Japanese saw themselves to be, and the non-American managers saw the Americans as even more open that they saw themselves to be.
4. The British managers were perceived as being more closed interpersonally than they saw themselves to be.
5. The Japanese global managers saw themselves as being only slightly closed interpersonally, but the members of all the other groups except the Singaporeans saw the Japanese as being very closed (the Singaporeans, who worked for the Japanese, thought they were only slightly closed).

In a variant of the preceding findings to explain differences across nationalities in perception and evaluation, Caligiuri and Day (2000) found that managers who enacted behaviors that were designed to reflect positively on their superiors were rated more highly by superiors of their own nationality than by superiors of other nationalities. Thus, trying to project an image of oneself as a solid performer is a culture-specific process that does not transfer across cultures in the performance appraisal context.

This state of affairs should not surprise anyone, but when it comes to conducting valid performance evaluations, perceptual biases among people from different cultures are no longer a topic of amusement or intellectual curiosity; such cross-cultural biases can threaten both the validity of performance evaluations and managers' careers. Consider the following "close call" that an American global manager experienced in France (Oddou & Mendenhall, 2000):

> In France, women are legally allowed to take six months off for having a baby. They are paid during that time but are not supposed to do any work related to their job[s]. This expatriate had two of three secretaries take maternity leave. . . . The American expatriate asked them to do some work at home, not really understanding the legalities of such a request. The French women could be fired from their job[s] for doing work at home. One of the women agreed to do it because she felt sorry for him. When the American's French boss found out one of these two secretaries was helping, he became very angry and intolerant of the American's actions. As a result, the American felt he was given a lower performance evaluation than he deserved. (p. 221)

Set the scene just described in your mind. One secretary is in the American's doghouse because she will not do any of her work at home. The other secretary is in the American's good graces but in the French manager's doghouse for breaking the law. The American is in the French manager's doghouse, but initially he does not know why. When he finds out why, he cannot get the French manager to understand his reasoning for why he did what he did. The French manager thinks that the American is insensitive and possibly incompetent.

What happened next? The American asked his former boss, another American, to intercede with the French superior. After a while, the French manager came to understand why the American had done what he did, and he modified his ratings and comments to make them more reasonable to the American. (What happened to the secretaries is unrecorded.) The point is that the French manager simply believed that the American manager should have been aware of French laws governing maternity leave, and this assumption led to problems for everyone involved. Such "war stories" are common among all managers on global assignments; they are part of the territory. What is important, however, is that if cross-cultural misunderstandings remain unresolved, they can negatively affect performance appraisals.

RESOLVING APPRAISAL DILEMMAS IN THE GLOBAL CONTEXT

Companies can do several things to lessen the likelihood that invalid performance appraisals will be conducted in their overseas operations. We have charted an approach for appraising global managers that is general enough to be applied across firms and industries yet specific enough to directly help the people doing evaluations.

What Should Be Evaluated?

Uncovering the key factors of success for an overseas position is not easy; there are no shortcuts. Only careful analysis can reveal these key factors. It is a given that the company wants to achieve profits or ROI, but external conditions may affect these figures making it difficult to determine how well the manager is doing when exchange rates fluctuate wildly, transfer pricing distorts profit pictures, and differences in accounting procedures affect financial reporting. Each company needs to define specifically what it means by success in Chile, Japan, the People's Republic of China, Great Britain, Nigeria, or any other country in which the company has subsidiaries or operations.

Obviously, business strategy largely dictates what is expected of a manager in a specific country. It would be foolish to focus heavily on profits as a criterion of success for a manager in the People's Republic of China, for example. If the company is in that country to make a lot of money in the short term, top management will be frustrated. More than likely, the company entered China hoping to build a presence there and eventually, over a period of many years, to position itself to tap that country's burgeoning market. Thus, the question becomes, What should a global manager be doing in Beijing to execute the company's strategy?

Superimposing what makes sense from a home perspective would be counterproductive. The keys to success in China are to develop close personal relationships with key government officials and to invest in training Chinese workers over a long period so that when government restraints on business dissolve, the company will be positioned for success. An important issue to resolve then is whether the firm's strategy in the country makes sense. If it does not, managers will be between a rock and a hard place, feeling compelled to give the home office what it demands while simultaneously expending energy on what they consider to be the real issues.

Even if the strategy does make sense and the evaluation criteria are clear to everyone, the ways to meet these criteria may need to be different in the host country than at home. For example, if output of production is an important criterion, the manager in Chile will need to spend most of his or her time on labor-management relations. In Chile, the key to keeping output on target may not be to focus on supply costs and inventories but simply to avoid strikes and get people to work on time. In locations where workforce stability cannot be taken for granted, spending time developing personal relationships with union leaders may make the difference between a plant being closed twice instead of 22 times during the manager's tenure (Oddou & Mendenhall, 2000).

How can a company fine-tune performance appraisal criteria to specific locations? Time simply must be spent on getting the facts. Executives must travel to the locations to observe, ask a lot of questions, and solicit insights from both current and past managers working in these countries. At the home office, outside experts from universities, consulting firms, and government agencies as well as other individuals who have lived and worked in countries where the company has operations should be brought in and thoroughly interrogated for information that could help uncover the key factors of success in the locations in question.

Ideally, a manager who has returned to the home office from an overseas site should be a permanent part of the team that updates performance criteria for overseas assignments. Reevaluating the criteria and their priority periodically will ensure that performance evaluation criteria reflect the reality of situations overseas (Oddou & Mendenhall, 2000). It may make sense to have repatriated global managers travel to their international sites every 3 to 6 months to see how the current manager is performing, to interview the manager, and to gain insight into the country's current business climate, the challenges facing the firm in that country, and any changes that are needed from a strategic or an operational perspective. In this manner, the criteria on which the manager is evaluated remain fluid, changing in response to the business and cultural climate.

In short, there needs to be interaction between managers who are currently on global assignments or managers who have been repatriated and HR and line managers who develop the criteria for performance appraisals of overseas managers. This approach has been called for by a variety of researchers, and some scholars are seeing a growing trend in favor of this approach (Suutari & Tahvanainen, 2002; Tahvanainen, 1998; Woods, 2003).

Who Should Conduct the Evaluations?

Gregersen et al. (1995, 1996) found that on average, firms had three evaluators for every international assignee; roughly 12% used only one rater, 43% used two raters, 20% used three raters, and 15% used four or more. By comparison, in purely domestic situations, these same companies used an average of six raters per employee.

Companies vary widely on who appraises managers on global assignments. Figure 7.2 was derived from the research findings of Gregersen et al. (1996) and

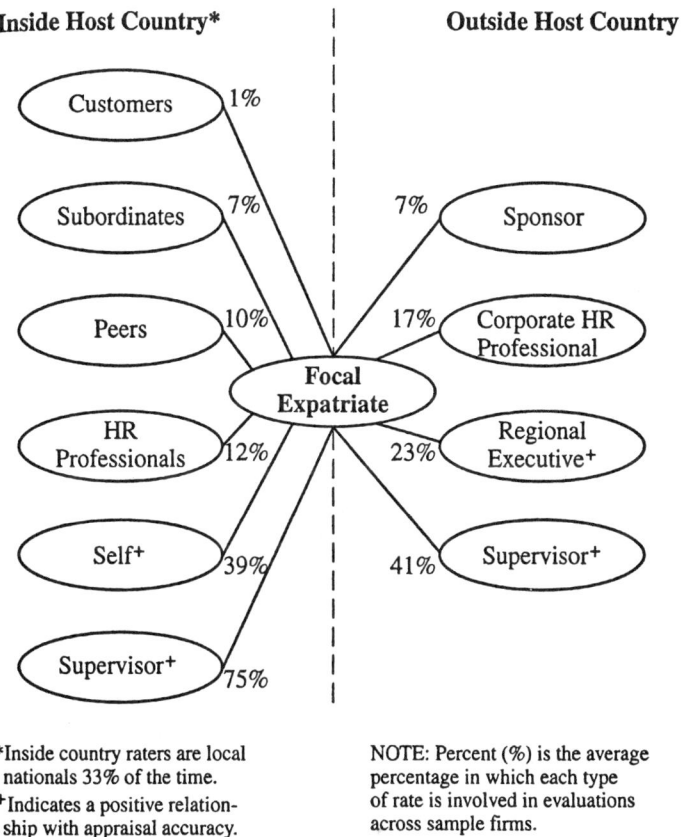

*Inside country raters are local nationals 33% of the time.
+Indicates a positive relationship with appraisal accuracy.

NOTE: Percent (%) is the average percentage in which each type of rate is involved in evaluations across sample firms.

FIG. 7.2. Raters of expatriate performance.

reflects the vast differences in approaches to performance appraisal in their sample of multinational companies. The permutations of who can and is used to evaluate performance are wide in scope, and not surprisingly, multinational companies take a variety of approaches. For example, one study found that U.S. managers on global assignments are most often evaluated by people from their home offices (Stening et al., 1981), whereas other studies have found that a majority of the raters are insiders (Gregersen et al., 1996). Other studies show a similar range of findings (see Woods, 2003).

We propose that leading-edge companies take more of a team approach to performance appraisals. This approach has correlated highly with whether respondents perceived the performance appraisal to be effective and fair (Gregersen et al., 1996). The following description captures this leading-edge approach:

- A team of organization members, led by a team coordinator, is assembled for the performance appraisal of the company's global managers. The team coordinator, often a senior human resource management executive, serves as the "hub" of the multiple evaluation process. This executive's task is to collect, analyze, and draft a "global manager evaluation report" that reflects feedback from the team members.

- The report is then forwarded to the individual who has direct-line authority over the global manager and to other senior executives and is filed at corporate headquarters in the human resource department.

- The hub manager coordinates the collection of feedback from the global manager's on-site superior (if the global manager is not the head of the overseas operation), peer managers of the global manager, subordinates of the global manager, clients of the global manager (if relevant), and the global manager himself or herself.

The members of the global manager's performance evaluation team and their relationships are illustrated in Fig. 7.3. Given the cross-cultural complexity of the global manager's assignment, it is necessary to have multiple raters. Ideally, these raters reflect the different constituencies of the global manager. Notice that home-office personnel make up a small minority of the team. Research shows that the most basic prerequisite for conducting a performance evaluation is that the rater has to have an adequate opportunity to observe the ratee's job performance over a reasonable period of time such as 6 months (Casio, 1986). By definition, someone from the home office is unlikely to qualify. Thus, the home office needs information regarding the quality of the global manager's performance from those who have had an adequate opportunity to observe the manager at work abroad.

It is dangerous to rely on one rater's evaluation of a global manager given possible rater bias stemming from cross-cultural issues. For example, if the global manager's immediate superior is of the same nationality as the global manager, the superior may feel that the manager is doing wonderfully. The reality, how-

ARE PEOPLE DOING THE RIGHT THINGS? 157

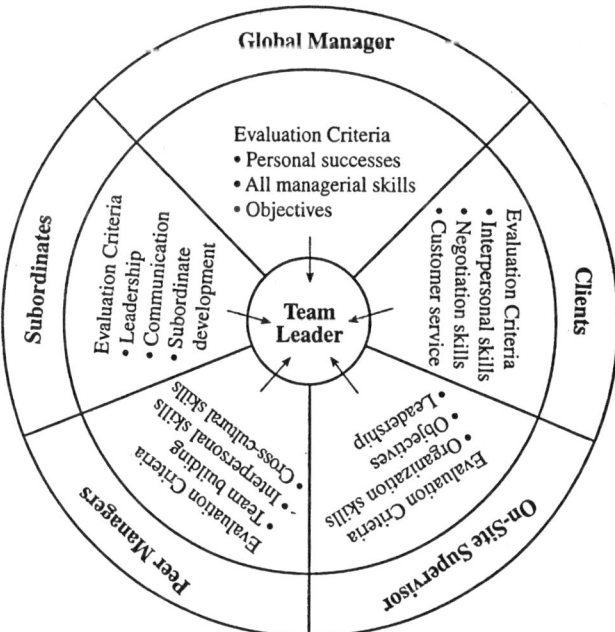

FIG. 7.3. Global manager evaluation wheel.

ever, may be that both individuals are poor performers because of their lack of familiarity with the local culture, the business climate, negotiation norms, and so on. An unwarranted "halo effect" may occur resulting in the manager receiving an inflated performance rating. Conversely, as illustrated in the example involving the global manager in France, an immediate superior who is a host-country national may be too tough on the global manager because of cross-cultural misunderstandings. Thus, relying solely on the rating of one person will not give a clear picture of a global manager's performance. The performance evaluation wheel in Fig. 7.3 reflects a more comprehensive approach to accessing feedback about managerial performance on global assignments.

On-Site Superiors. These individuals are in the best position to report on the relation between a global manager's job performance and the organizational goals and objectives of the foreign unit and to observe the manager's performance regarding important tasks, projects, and organizational concerns.

Peer Managers. Obtaining feedback from peer managers in the overseas setting, of various nationalities if relevant, will reflect to some degree how well a global manager works with others to accomplish the organization's objectives. The responses of these managers will also reflect the degree to which a global manager possesses managerial skills, interpersonal skills, cross-cultural skills, and multicultural team-building skills.

Subordinates. Ratings from a global manager's subordinates, especially if they are mostly host-country employees, will counteract halo-effect ratings by an immediate superior who is of the same nationality as the manager. Feedback from subordinates will provide some insight into the manager's cross-cultural managerial skills, communication skills, leadership abilities, and the degree to which the manager can work effectively among individuals whose values, work style, and other cultural characteristics may be different from the manager's.

Clients. A global manager's job description usually includes representing the firm to external institutions with the goal of enhancing the company's public image, generating business, influencing government decisions related to the company and its industry, fostering alliances, and generally gaining favor in the larger community. The only way to ascertain a manager's success in achieving this goal is to question a sample of clients. Clients will not be able to provide insight into an individual's in-house managerial skills, but they will be able to provide useful feedback about the manager's negotiation skills, achievement of "figurehead" duties, interpersonal skills, sales skills, and alliance-building skills.

Self. Although the opportunity for rater bias exists in self-ratings, international assignees should be given a chance to express their accomplishments, frustrations, progress, and challenges since the previous appraisal was conducted. Personal points of view are an important piece of the performance puzzle. Also, because global managers know that their input is only one part of the appraisal network, they will be motivated to write reports that do not wildly exaggerate their performance.

When Should Appraisals Be Done?

Much research has been conducted on when appraisals should be done, and the results seem fairly clear. "Research over the past twenty years has indicated that once or twice a year is far too infrequent. Considerable difficulties face a rater who is asked to remember what several employees did over the previous six to twelve months.... People often forget the details of what they have observed and they reconstruct the details on the basis of their existing mental categories" (Casio, 1986, pp. 302–303). Managers tend to evaluate performance in a 6-month period based on what happened in the last 6 weeks rather than across the entire time period.

The dilemma is that for the sake of timely and consistent performance appraisals, effort is drained from managers who could be focusing their energies on activities that enhance profits—after all, organizations do not exist so that elaborate performance appraisals can take place. A balance should be achieved between the need for timely evaluations and the need to spend time on more task-oriented activities.

Given these constraints, we propose that team leaders prepare reports on global managers every 6 months. That does not necessarily mean that all the input should be gathered only twice a year. Rather, performance appraisals should be conducted in the following time frames and feedback obtained from the following sources.

On-Site Superiors. Managers on global assignments should be appraised not after specified time periods but after the completion of significant projects, tasks, or other organizational milestones. This method enables the manager's superior to conduct the evaluation in the context of a specific task, thus avoiding the likelihood of rater bias due to a recency effect. Reviews are then forwarded to the team leader and are filed for later use when the team leader formally writes the global manager evaluation report.

Peer Managers. Peer managers should be asked to evaluate global managers once every 6 months. By nature, their interactions with these managers will be sporadic—sometimes heavy and sometimes light—so 6-month intervals are necessary to build up a mental database on which to base their reports.

Subordinates. Subordinates should be asked to evaluate the global manager's activities at the same time as the on-site superior is preparing his or her evaluation—that is, after major projects are completed. This approach prompts subordinates to evaluate the manager with respect to direct task-oriented issues rather than on general attitudes about the manager.

Clients. Clients should be accessed once a year. Clients are busy people who do not want to be asked to fill out forms every 3 months. What is the proper way to approach clients and ask them to be part of an evaluation process? The global manager, the on-site superior, the regional superior, or executives at corporate headquarters may want to generate lists of clients who could be contacted. Feedback from these clients could be gathered formally (via phone interviews and surveys) or informally (over dinner)—whatever makes sense in terms of the clients' personalities and cultural preferences. If contacting clients personally is seen as too intrusive, companies may want to ask external observers to evaluate the manager's dealings with clients. These observers could be in-house managers or consultants.

Linking Patterns of Globalization and Strategy Orientation to Appraisal

Again, we emphasize the importance of seeking a fit among a company's people-management practices and its stage of globalization and competitive strategy. In addition, the company's five people-management functions must support and reinforce one another if each function is to be performed successfully. For example, little long-term benefit is derived from carefully organizing an elaborate performance appraisal system for global managers and then, at the completion of the international assignment, placing these individuals in positions that do not make use of their international skills. The appraising function allows good management development and career planning to take place; it is an important link among functions in the people-management process.

Throughout this chapter, we have focused on ideas that will enhance the effectiveness of the appraisal system for global managers. All companies have different strategies, however, and are at different stages in the globalization process. We would like to discuss the appraisal function briefly in this light. With respect to performance appraisals, the following two general principles apply:

1. The more a company's competitive strategy moves toward differentiation, the more comprehensive and complex the performance appraisal system needs to be. Even at the export stage of globalization, if a firm has a differentiation strategic orientation, global managers must have above-average cross-cultural skills to be effective. They need to understand the culture well because products or services are designed to be culturally compatible with the consumers' preferences.

2. The more a company's stage of globalization moves away from export and toward more integrated stages (MNC and global), the more comprehensive and complex the performance appraisal system needs to be. As companies become more globally integrated, managers find themselves not only responsible for "hard" organizational performance objectives (e.g., profits, ROI, market share, etc.) but also for "soft" organizational performance objectives (e.g., socializing host-country managers into the firm's corporate culture, coordinating and balancing the home office's and the subsidiary's concerns, representing the firm to external stakeholders, etc.). Thus, the criteria for success increase in number and magnitude, making the team approach to performance appraisal necessary.

SUMMARY

No matter how many managers a firm has overseas, they have to be evaluated in a way that makes sense in the context of each one's overseas situation, the firm's strategy for that overseas unit, and the stage of globalization the firm is in. The overarching principle to remember is that no matter what the context, the global manager should be evaluated on criteria that are truly important for success overseas and not on criteria that are ancillary to success. This hurdle is difficult to overcome when conducting international performance appraisals, but tackling it will yield more effective and successful global managers.

CHAPTER

8

Rewarding: Compensation and Pay

Arguably, no aspect of international assignments gets more time, attention, and money spent on it than the topic of compensation. Individual managers have a strong personal interest in the topic as do companies—although the interests are not always 100% aligned. Based on extensive discussions with executives around the world and several surveys, we have identified three common and important objectives of international assignment compensation systems:

1. Enhance the mobility of top talent. As we discussed earlier in this book, global companies understand that company culture, know-how, market intelligence, and other important intangibles require human, face-to-face interaction for maximum dissemination and inculcation. Global firms need people to move around. At a minimum, compensation systems should not inhibit mobility, and at a maximum, they should enhance it.

2. Foster fairness and equity. For those making international moves, the compensation package before, during, and after relocation needs to be seen as fair and equitable; we call this "within-person fairness." In addition, as people move across geographic borders, they contrast and compare their compensation with that of other employees both inside and outside the company. Compensation systems also need to be seen as fair and equitable across geographies; we call this "across-person fairness."

3. Control costs. Finally, while achieving the first two objectives, international assignment compensation systems also need to control costs in dollars spent on compensation and in dollars spent administrating the compensation system.

As discussed in the first of two "From the Front Lines" in this chapter, many complex employment-related questions need to be answered the moment a company decides to send an employee abroad.

FROM THE FRONT LINES:

"Baker & McKenzie's Benefits and Compensation Options for International Assignees"
by Kerry Weinger, Partner, Baker & McKenzie

Any time Baker & McKenzie works with an employee going on an international assignment, several employment, benefits, and compensation law questions must be considered. For example, there are numerous ways to structure the employee's assignment from an employment law point of view. The most common approaches are to maintain the person as an employee of the home-country employer and "second" him or her to the host location, employ the individual by the host-location employer, employ the individual by a special services company (such as a global employment company) and have the global employment company second him or her to the host location, or make the individual an employee of both the home-location employer and the host-location employer. The employment structure will affect benefit arrangements such as the employee's ability to continue participation in a company's home-country plans.

The following are some of the other critical issues to be addressed:

- Where will the expatriate be subject to taxation?
- What provisions should be contained in the expatriate's international assignment agreement?
- What are the employer's and the expatriate's legal responsibilities?
- What are the home- and host-country tax consequences of the assignment?
- What planning is available to minimize the employee's and the employer's tax liabilities?
- What are the home and host location's Social Security consequences?
- What immigration rules apply?
- What are the financial, human resource, and legal consequences of terminating the assignment before its expected term?
- What are the ramifications under the local labor laws of terminating an employee or changing the employee's compensation and benefits?

As with other areas of the international assignment process, by anticipating legal problems before the assignment, the employer will be better able to protect itself from unanticipated problems and liabilities later on.

PROBLEM OF REWARDING A WHILE HOPING FOR B

One of the problems that compensation systems create is what is commonly referred to as "rewarding A while hoping for B" (Stroh, Northcraft, & Neale, 2002). Essentially, this phrase captures the problem of rewarding one behavior while hoping that the person will actually exhibit another. For example, professors hope that students will learn classroom material and expand their understanding, but they often reward only the ability to regurgitate memorized facts and figures. In business, employees are often quick to point out that although their bosses hope they will show initiative and solve problems, bosses often reward employees for simply following orders and sticking closely to standard operating procedures.

Because the problem of rewarding A and hoping for B is so common, examples abound. Two of the reasons that are offered for this pervasive problem are as follows: (a) objectives often are not clear or are not clearly understood, and (b) even when objectives are clearly understood at the outset, the means of achieving the objectives eventually supplant the original objectives and become objectives themselves; reward systems for international assignees are just as vulnerable to this phenomenon as any other reward system.

ENHANCE THE MOBILITY OF TOP TALENT

One of the primary objectives of reward systems is to attract and retain high-quality people in international assignments. Although some employees may see the positive and appealing aspects of living in a different country and culture, living abroad is fraught with numerous uncertainties and negative consequences. Leaving family, friends, familiar and comfortable living conditions, education and health care facilities, entertainment and recreation opportunities, favored foods and shopping areas, and so on is not something most people want to do. Consequently, money is often used as a means of "buying off" the loss of what is left behind or as a means of simply buying employees or paying so partners will go along. As one manager in Jakarta pointed out, "The normal grocery store here doesn't carry Wheatees, but with enough money, I can get my Wheatees."

Consequently, although firms may hope that high-quality people will be attracted to international assignments and hope that these individuals will adjust effectively to their new culture, many of the rewards of accepting a global assignment (which we explore in detail later in the chapter) often encourage employees and their families to transport or purchase a lifestyle similar to the one they had back home and not make much effort to adjust to the new culture and country. In other words, company compensation systems can hope that they will attract high-quality people to go abroad and learn about international competitors, customers, cultures, and so on; however, these systems also create incentives and means for people to live in expatriate-dominated neighborhoods, buy imported foods from home, send their kids to international rather than local schools, and hang out at the American Club (or the equivalent for other nationalities) and socialize not with local nationals but with other managers on global assignments from back home. In other words, if companies are not careful, they end up rewarding A while hoping for B.

FOSTER FAIRNESS AND EQUITY

Ideally, international reward systems are designed to enhance feelings of equity, which should not be confused with feelings of equality. Most organizations are not built on the principle of equality—that is why firms are generally structured like pyramids with various levels of hierarchy and why financial compensation almost always increases as a person moves up the pyramid. The notion of equity assumes that people determine the ratio between what they "put in" or contribute and what they "get out" or receive. This means that employees at the same organizational level performing at virtually the same level of performance would expect quite similar rewards. In the international context, this means that there should not be significant differences in pay between managers from different countries who are assigned to the same location and performing equally well at similar jobs. Organizations also should strive to avoid significant differences in pay for managers performing equally well at the same job in different locations.

Unfortunately, the financial incentives that are used to attract high-quality people to international assignments often create a strong sense of inequity. Local nationals just one level below global managers frequently see large gaps between what they contribute and the rewards they receive and what global managers contribute and the rewards they receive. Many of the rewards that global managers are given are lost if they are not used, which only encourages them to use these rewards or lose them. In addition, poorly designed reward systems can create inequity between employees originating from different countries but assigned to the same location. For example, a manager in Japan who heads up the local office for a large U.S. multinational commented, "I have a German expatriate working two levels below me with a compensation package equal to mine and twice as large as a British expatriate one level above him." Firms may hope for equity, but many of their reward systems encourage the maximization of inequity. Consider the following comment from the spouse of an American expatriate.

> I feel this firm does not really care about its people and their families as well as it should. Everything is done on the "cheap," which makes adjusting to the new country so much harder, which in turn makes their employees even more stressed out. Comparing this assignment to other major New York banks' employees, our benefits were extremely standard and embarrassing! For example, the housing allow-

ances of many people we met were 25% higher than ours. Home-leave airfares were only "economy" in our company, while other firms were "business class." Significantly higher bonuses were given by other firms. Our firm has lost and will continue to lose a great many excellent people should they continue to deal in this way. By the way, my husband left the bank for these reasons since our repatriation.

COST CONTROL

Although first movers may enjoy a general business advantage, when it comes to compensation, there is more often a first-mover penalty. We do not have to go back many years to find times when international business as a percentage of a company's total revenues was small, meaning international assignments and assignees were not so important. In 1980, when Jack Welch took over as CEO of GE, only 10% of the company's revenues came from outside the United States. Because international operations were of little importance, international assignees were out of sight and out of mind.

When it came time to come home and capture the anticipated career payoff, countless employees were severely disappointed. As a consequence, many companies had to provide a variety of incentives to get reasonably qualified people to move abroad. Once these incentives were in place and other firms benchmarked (copied) the practices, it was hard to remove them. The first firm to try would hear resounding complaints from its employees about how such and such a firm provided the threatened benefit, allowance, or perk.

As firms such as GE and others found international sales a key to continued growth, the importance and cost of international assignments grew and the number of assignees increased. Cost control became more and more important when managers living abroad cost a company two to three times what it was spending on a high-quality employee back home. It was hard to take away benefits that had been offered previously and were common rewards for global managers. Some companies simply reduced the number of employees going abroad, only to find that coordinating global operations was more difficult, creating and sustaining a cohesive culture was more elusive, and finding executives with the needed global mindset and experience was more challenging. Thus, whereas many companies hoped for reasonable cost control and enhanced globalization, they rewarded compensation managers for lowering the absolute costs spent on employees working abroad at the expense of achieving their first objective.

Today, cost control is viewed more in terms of cost per global employee than in terms of total costs spent on international locations. As a consequence, there is considerable emphasis on administrative costs. How can the administrative costs overall and per expatriate be reduced? How can efficiencies be captured? What approaches and practices can be used to lower administrative costs without hurting the chance to meet mobility and equity objectives? In addition, as we discuss, companies are looking at approaches and specific practices and policies that can bring down the cost per global employee while not compromising the mobility and equity objectives.

BASIC COMPENSATION APPROACHES

In our experience, approaches used to compensate employees on international assignments vary more in "how" they try to achieve the objectives discussed previously than in "what" their objectives are. Regardless of approach, companies recognize that their best and brightest employees will not accept international assignments if they do not feel they are undertaking a "good deal." Companies also recognize that a large part (although not 100%) of what gets factored into this decision is money. This is why most compensation approaches to international assignments have as their fundamental objective to ensure that individuals are not hurt economically by taking the assignment. In addition, experienced companies know that as long as employees are overseas, they will compare compensation packages against other packages both within and outside the company. If there are too many inconsistencies, special deals, and the like, employees end up feeling that they were treated unfairly. If the feelings of inequity and unfairness are strong and widespread enough, the compensation system ends up hurting rather than enhancing the desired level of mobility. Finally, all of the basic approaches recognize that if left unchecked, compensation costs can easily spin out of control. On this objective as well, compensation systems differ more by how they try to control costs than by whether they think costs should be controlled.

BALANCE SHEET APPROACH

The balance sheet approach is arguably the dominant compensation system used in multinational companies. The approach has been used for more than 20 years, yet nearly two of every three companies still use a form of this approach to compensation. The basic objective of the balance sheet approach is to ensure that employees can maintain a standard of living in the country of assignment similar to that which they enjoy in their home country. This keeps the shock from changes in standard of living relatively small. Figure 8.1 provides a rough illustration of how this approach is designed.

The basic premise of the balance sheet approach is that companies use a set of allowances or differentials to maintain employees' purchasing power while on international assignments. Fundamentally, assignees should neither benefit nor be hurt economically as a result of relocating.

COMPENSATION AND PAY

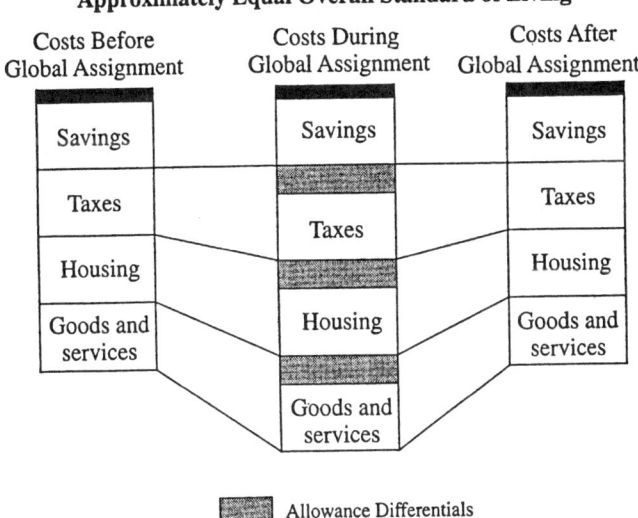

FIG. 8.1. Balance sheet.

Taxes

Most experienced managers we have spoken with label taxes as one of the most important and complex aspects of any international assignment compensation system. Within the balance sheet approach, there are two general tactics for dealing with taxes.

The most common approach is usually termed *tax equalization*. With this approach, a "hypothetical tax" is estimated based on what global employees would have paid at home. That money is then subtracted from their salaries. The firm pays all actual taxes that employees owe in the home and in the host country.

There are several advantages to tax equalization. First, subtracting the hypothetical home-country tax from the employee's total salary reduces actual taxable income. However, in 1990, Organization Resources Counselors (ORC) found differences in how this hypothetical tax was computed. Roughly 25% of U.S. firms estimate it on base salary alone; 54% estimate it on base salary plus bonus; and 8% estimate it on base salary, bonus, and all premiums.

The benefit to the firm of tax equalization can be quite substantial, especially if employees originate from high-tax countries. This is probably why nearly 9 in 10 U.S. firms use tax equalization. Second, it reduces problems on repatriation. For example, if an employee from the United States were sent to Saudi Arabia (which has virtually no personal income tax) and was allowed to keep the tax windfall, the employee would experience a rather significant shock on repatriation. The person could easily experience a drop in disposable income of 30% to 50%.

Third, making adjustments through tax equalization for differences in income tax policies in different countries provides motivation to employees from low-tax countries such as Saudi Arabia to relocate to high-tax countries such as Sweden. Fourth, if the majority of the firm's employees sent on international assignments originate from countries with relatively high personal income tax, the firm is likely to receive a net benefit (or cost reduction) from tax equalization as a result of the differential tax rates.

The second approach to handling taxes is commonly referred to as *tax protection*. Under this policy, firms reimburse employees for taxes they have to pay that are in excess of taxes they would have paid if they had remained in their country of origin. There are several things to keep in mind. First, in some countries, some monetary incentives, discussed in more detail later in the chapter, add to employees' taxable income, making it higher than it would be if they had remained home. Second, in some countries, the extra money reimbursed to compensate for these additional taxes is itself taxable income. Thus, a cycle is set in motion whereby additional taxable income is generated along with the need to provide further reimbursement. Carried to the extreme, this can become a never-ending cycle of compensation and tax reimbursement. Consequently, most firms limit this cycle to one or two rounds. The other major issue to keep in mind is that when employees relocate from higher tax countries to lower tax countries for their assignments, they may owe less taxes than they would if they had stayed home. In this case, firms must decide if they will allow employees to keep the difference or require them to reimburse the firm. Most firms in this situation find it difficult to require employees to reimburse the firm for the difference.

Housing

Although housing may not be quite as complex an issue as taxes, it is of similar importance in the minds of both individual international assignees and companies. With the balance sheet approach, there are two different tactics for trying to ensure that employees pay no more or less for housing than they did at home. With the first approach, a housing norm for the home country is computed and that amount is deducted from the employee's salary. All approved housing expenses in the host country are then paid directly by the company. With the second approach, the difference between the cost of comparable housing in the host country and the home country (i.e., housing differential) is paid directly to the employee. The employee then has two options: (a) to live in less expensive housing than the home-plus-housing differential and pocket the difference or (b) to live in more expensive housing and pay the added cost.

The housing norm approach works relatively well under several important conditions. First, because the objective is to keep the standard of living constant before, during, and after the international assignment, it assumes that individuals return to their home country after they complete the international assignment.

This is the case for many firms. However, some firms have a system in which a cadre of career internationalists moves from one international assignment to another. In this case, it becomes difficult to determine (a) the employees' home country and (b) the rationale for why their standard of living abroad should be equal to that often arbitrary country.

Second, this approach works best when most of the firm's employees on international assignments originate from the same country. If expatriates are from different countries with different standards of living and yet work together in the same country, the balance sheet approach to housing actually serves to highlight the imbalance or unequal standards of living these individuals enjoy (or suffer). Employees with equal responsibilities may find it difficult to accept unequal housing differentials due primarily to accidents of birth.

About 20% of U.S. firms have instituted a modified approach to housing differentials. Rather than strictly computing differentials based on home and host-country costs, they develop a housing standard based on job level. The company may determine that a vice president on assignment in Hong Kong should pay rent of HK$45,000 per month. This standard would ensure that employees in Hong Kong who were of the same grade and salary in the company had similar housing. This approach tends to work best when a firm has employees from a variety of different countries in a given locale and when the firm emphasizes across-person equity over within-person equity and host-country norms.

As another example, consider the case of a financial service firm that had three vice presidents working in Hong Kong. One was from Tokyo, one was from San Francisco, and one was from Salt Lake City. For HK$45,000, they could buy an apartment in Hong Kong that was much bigger than the executive from Tokyo had at home. It would be about the same size as what the executive from San Francisco enjoyed at home but only half the size of the house on nearly an acre of land that the executive from Salt Lake City had prior to the assignment. They would all feel reasonably good that at similar levels in the company, they had similar levels of housing, but had there not been an emphasis on host-country norms in calculating housing allowances, the executive from Salt Lake City would have felt "cheated."

Third, this method is most effective as long as reliable and detailed figures exist to calculate and compare standards of living. For example, consider the case of a biotech firm with headquarters in San Francisco, California, and regional headquarters in Geneva, Switzerland. Recently, the firm transferred one of its leading R&D managers, Randy Keller (married with two small children), to Switzerland from Los Angeles. The director of corporate personnel made some inquiries about the annual cost to rent a three-bedroom home or apartment in Geneva. He found out that the annual rental cost would be approximately $60,000. The director knew that the annual cost to own a three-bedroom home in San Francisco was about $50,000. This translated to a $10,000 housing adjustment that needed to be made in favor of the employee. Instead of these figures, the director could have easily relied on national averages for Switzerland and the United States in determining Randy Keller's housing allowance. The point is that housing costs vary not just by country but by region, city, and even neighborhood. For example, Randy Keller's annual cost for his three-bedroom home in Los Angeles were $45,000 ($5,000 lower than in San Francisco). The average annual rental cost for a three-bedroom apartment in Switzerland was $50,000 or $10,000 less than in Geneva.

From this simple example, it should be clear that in a firm's effort to ensure that employees pay no more or less for comparable housing and thereby enhance international mobility and a sense of fairness, a company may in fact increase the complexity and cost of administering its housing policy. It doesn't take a very large company with many employees in just a few countries before the cost of acquiring detailed housing costs becomes nontrivial. Although firms such as ORC are able and happy to provide these figures, they come at a cost. The more details a firm has to have, the more it will have to pay to get detailed figures.

Goods and Services

Once taxes are equalized and housing differentials are computed, the balance sheet approach calls for the determination of goods and services differentials. In many companies, these goods and services are termed cost-of-living allowances (COLAs). COLAs are utilized in recognition that the costs of equivalent standards of living vary by country. Approximately 9 out of 10 U.S. firms pay such allowances. Procter & Gamble is one such company. Its COLA program is discussed in the second "From the Front Lines" in this chapter.

Essentially, firms must themselves or through consultants assess the cost of a comparable "basket of goods" in the country of assignment relative to some comparison country (usually the country of origin). If goods and services cost more in the country of assignment than in the comparison country, a cost-of-living adjustment is made. If the cost of living in the country of assignment is lower than in the comparison country, adjustments are usually not made; very few U.S. firms make negative adjustments.

The general method of making COLA calculations can be illustrated with the following example. Suppose Kay While, a senior manager with a large accounting firm, was being transferred from Boston to Tokyo. The accounting firm contacted a consulting firm specializing in overseas COLA calculations. The consulting firm determined that it costs a single individual without a family 65% more to live in Tokyo than in Boston. Next, the consulting firm estimated that 30% of Kay's $100,000 salary was for goods (food, clothing, transportation, recreation, personal care, medical care, etc. but excluding housing). Accordingly, the consulting firm determined the COLA for Kay While in Tokyo to be $19,500. This COLA or goods and services differential is generally given tax free; if any additional taxes are associated with the differential, the company bears those taxes.

> **FROM THE FRONT LINES:**
>
> ### "Procter & Gamble's Program to Bring Cost-of-Living Differentials Into Line"
> by Marie Howard, Senior Manager, Global Delivery, Relocation & Expatriate Services, Procter & Gamble
>
> For Procter & Gamble's globalization strategy to succeed, the company must be able to ensure that the best qualified employees will be willing to accept international assignments. One of the issues standing in the way of employees at some companies is the relatively high cost of living in many countries. At Procter & Gamble (P&G), we keep our international assignees on a home-based compensation plan and pay allowances for the new location, housing and utilities, and cost of living.
>
> To provide our international assignees with a truly home-based compensation system required creativity when designing our plan. We knew that by maintaining a home-based plan, assignees with lower home-based salaries would be unable to maintain a comparable lifestyle when they were transferred to a location where goods and services, even with a cost-of-living adjustment, were significantly more expensive than at home.
>
> As a way of avoiding this problem, P&G offers a supplementary allowance to offset the higher cost of goods and services in some countries. Based on the employee's family size and base home salary, we calculate the cost-of-living differential between the cost of goods and services at home and in the host country and compare that to the amount of income the employee has to spend on goods and services. Based on this differential, we determine a supplementary allowance, which international assignees are given each month. This encourages employees with lower home-based salaries to transfer to higher cost locations and enables employees and their families to maintain a comparable visible lifestyle to that of their peers.

Most COLA differentials are subject to review to ensure their appropriateness. Changes in prices in either the home country or the host country as well as changes in exchange rates can significantly affect the appropriateness of a goods and services differential.

The timing of inflation reviews and the rate of inflation can have a particularly important effect on the COLA differential. For example, if the annual inflation rate were 120% in, say, Thailand, then the frequency and timing of COLA calculations could have a significant impact. Suppose that company A evaluated inflation every 3 months whereas company B evaluated it every month. Every 3 months, employees of both firms would have approximately 30% more baht added to their COLA. However, the purchasing power of employees in company B would have been much better protected. For example, if one employee from each firm received 300,000 baht in January, both employees would receive approximately 390,000 baht in April. However, whereas the employee in firm A would have received 300,000 baht in both February and March, the employee in firm B would have received 330,000 in February and 360,000 in March or a total of 90,000 more over the 3 months than the employee in firm A.

In addition, COLAs are subject to changes in exchange rates. Today, foreign exchange rates can fluctuate considerably. For example, in 2003, the dollar depreciated against the euro from $0.89 to the euro to $1.18. This is more than a 30% change. To illustrate the importance of exchange rate fluctuations, simply consider Maria Spyridakis who moved from Boston to Paris. Maria was given a goods and service differential of US$5,000. At the beginning of her assignment, that differential translated into 5,617.98 euro. By May 2003, the same differential translated into only 4,237.29 euro, a difference of 1,380.69 euro. From May through August, the dollar rebounded and appreciated to 1.09—a 7.5% change. At that point, Maria's goods and services differential translated into 4,587.16 euro, which was 349.87 euro better than in May but 1,030.82 euro worse than at the start of her assignment.

One important issue for firms to consider is that official exchange rates are not always accurate reflections of unofficial or real exchange rates. Especially in developing countries where the national currency is not freely traded on world exchange rate markets, governments have a tendency to overvalue their currency. Thus, although the dollar may officially buy only 8 Renminbi (RMB) in China, it may buy 10 RMB unofficially. In particular, the combination of high rates of inflation and slow adjustments in official exchange rates can have significant and negative impacts on true purchasing power. Failure on the part of firms to recognize these issues can often lead to dissatisfied employees.

BALANCE SHEET APPROACH: PROS AND CONS

The balance sheet approach is the most widely used because it has several advantages, but it is also not without disadvantages. The primary advantage is that it treats employees within a firm equitably regardless of their nationality by ensuring each employee protection relative to his or her home-country standard of living. This leads to the second major advantage: that the approach tends to facilitate international mobility because there are no financial disincentives for leaving home and taking an international assignment. Uncertainty is somewhat reduced, which also encourages mobility because the company bears any risk associated with unpredictable rates of inflation and currencies.

Among the downsides of the balance sheet approach is that the administrative burdens and costs can grow as the size and complexity of the firm increases. Obtaining the data needed to compute the differentials can also be costly. In addition, to the extent that the approach is successful in getting employees to accept international assignments, the chances increase that employees will see disparities between their home-country compensation levels and the levels of employees working in other parts of the world. This is not a hypothetical consequence but a real one. For example, in Tokyo, we heard complaints from a British general manager who reported to the country manager about a German expatriate essentially two levels below whose total compensation was higher than his was. When this German manager resided in Germany and the British manager resided near London, the differences still existed but were not salient. When they worked together in Tokyo, the differences in total compensation became clear because the balance sheet approach had been used to calculate COLA differentials.

Finally, this highlighting of differences is not restricted to comparisons between employees from different home countries living in the same locale. Comparisons can also take place between local managers and managers from the home office. Whereas local managers simply have to cope with local costs and bear the burden of inflation and currency fluctuations, outsiders receive differentials and adjustments that remove a lot of the risk for them.

PARENT-COUNTRY EQUIVALENCY APPROACH

This approach is in many ways quite simple and straightforward. Compensation for all expatriate employees (or all expatriate employees above a certain level) is based on market rates in the country where the company has its headquarters. For example, a U.S.-based multinational chemical company would pay all its international employees the same salary and give them the same benefits as employees in the U.S. regardless of their country of origin. This approach reduces the inequality often experienced when employees from different countries of origin work in the same locale.

The effectiveness of this approach is also subject to several important conditions. First, it works best when the parent firm's home country has relatively high wage and standard-of-living levels. Quite simply, this is because it is easier to convince individuals to accept pay scales and standards of living greater than they enjoy at home. The opposite is not an easy sell.

Second, this approach works best when the employees affected move from one international assignment to another. This reduces the negative impact that the loss of cost-of-living adjustments creates when individuals return home to countries with lower pay scales and living standards.

Third, this approach works best when the company has a relatively small number of managers (no more than 50) on international assignments or a cadre of career internationalists. Paying parent-country wages and ensuring parent-country living standards, especially when they are high relative to world standards, for a large group of managers working overseas can become very costly. Also, when large groups of managers are trotting around the globe receiving relatively high wages and standard-of-living differentials, local managers and employees are likely to feel they are being treated inequitably. This can hurt a parent-company manager's ability to work effectively with local staff.

The parent-country equivalency approach is relatively simple to administer and therefore often has lower administrative costs than the balance sheet approach because there is only one home country (i.e., the country where the company's headquarters is located) used to compute salaries and assignment differentials.

REGIONAL/COMPOSITE MARKET APPROACH

Some companies have found regional market or composite market approaches attractive. Although different terms are used for these approaches, they are similar. Both involve calculating an average salary and standard-of-living index for a set of countries. The term *regional approach* is generally used when the countries are geographically proximate (e.g., all in Central America, the Middle East, Scandinavia, Southeast Asia, etc.). The term *composite market* is used when the set of countries is based not on their geographic location but on some other criteria, such as their market importance to the firm. This approach is often less costly than the parent-country approach because it allows the firm to arbitrage the difference in wages and cost of living between the parent country and other countries.

Like the other approaches discussed in this chapter, this approach works best under certain conditions. First, ideally, the managers on assignment should come from and stay within regional or market boundaries. The more employees move between regions or markets, the greater the chance for feelings of inequity to emerge.

Second, this approach works best when region or market boundaries are drawn around countries that have wage and living standards that are more similar than different. For example, drawing a boundary around Canada, the United States, and Mexico and calling it the North American region would most likely create feelings of inequity because of the significant differences in wages and standards of living between Mexico and the other two countries.

LOCAL NATIONAL APPROACH

The fourth major approach to compensation for employees on global assignments starts from the premise that there is nothing special about being sent on an international assignment, and therefore, the firm should not have to provide special compensation or benefit packages. Accordingly, all employees, regardless of where they are working, are paid according to local market rates. A slightly less absolute version of this approach involves providing some allowances for the 1st year or two and then dropping them and treating transferred employees as local nationals. In another slight variation on this approach, the company pays a "one-time premium" at the beginning of the assignment or sometimes at both the beginning and the end. The obvious implication of this basic approach is that high-quality employees will accept international assignments only to countries where the wage structures and standards of living are higher than those of the country of origin. Enticing employees to make the opposite move is obviously difficult.

This approach can work under some fairly restrictive conditions. First, in firms that take a multidomestic strategic approach to business in general and that do not need much information transferred into or out of countries, there is relatively little need for managers to work abroad. Second, if the firm is expanding into countries that are more rather than less developed compared to the country where the parent firm or the most desirable employees are located, then the local national approach can work.

ALLOWANCES AND INCENTIVES

Although most basic international assignment compensation approaches, including the balance sheet approach, might have as their premise to avoid hurting employees financially, the reality is that many companies offer allowances and incentives to make taking an international assignment more lucrative than working at home. Although the cost-control objective in many companies is starting to get more weight so that some allowances and incentives are disappearing, the reality is that most companies are not populated with employees who see the world as their home. Many employees grow up in a given country with a specific native language, customs, and culture and are reluctant to leave family, friends, education facilities, medical systems, and so on with which they are comfortable and for which they have an affinity or love. As a consequence, although the overall trend is toward cost control and reducing allowances and incentives, most companies still find that they need to provide incentives to entice employees to move abroad.

FOREIGN SERVICE PREMIUMS (FSP)

An FSP is simply money that is paid in recognition of the employee and family's willingness and sacrifice in accepting an international assignment. The compensation is given in exchange for the inconvenience of living in a foreign land with alien customs, food, weather, transportation systems, education, shopping, and health care facilities and for leaving family, friends, and familiar surroundings back home.

The percentage of firms paying FSPs, sometimes called a mobility or international premium, declined from about 78% in 1990 to approximately 62% 10 years later. In most firms that pay this premium, employees receive a percentage, generally between 10% and 15%, of their base pay. Some firms set a maximum of, say, $50,000.

HARDSHIP OR SITE ALLOWANCES

If an FSP is paid, a hardship allowance is paid in addition in recognition that there are particularly difficult aspects of the country or site where the employee is being relocated. Hardships may include physical isolation, climate extremes, political instability and risk, or a poor standard of living due to such things as inadequate housing, education, health care, or food. Most firms pay hardship allowances, but the list of countries for which the allowance is applicable has changed and gotten shorter over time. For example, virtually no firms classify Singapore as a hardship post anymore, although a number of firms did just 10 years ago.

In identifying and evaluating hardships and determining a "fair" premium, most firms take a twofold approach. Many look at the competitive practices of similar firms that have managers in the countries or locations in question. If Motorola sees that GE offers a hardship allowance for Wuhan, China (but not for Shanghai), compensation managers will carefully factor that into their decision about whether to give a hardship allowance for relocation to Wuhan. Firms also use internal market tests. If desired employees simply won't go to Wuhan without a hardship allowance, then mobility objectives may trump cost control.

DANGER ALLOWANCES

Danger allowances are similar to hardship allowances except that they are paid in addition to hardship allowances for working in a hostile environment. These payments tend to be concentrated in certain industries such as oil and gas in which the location of raw materials (not economic development or political stability) determines where these companies choose to operate. Although an oil and gas company might feel it cannot pass up an opportunity in a civil war-torn country in Africa, it is likely to find that it has to pay danger allowances to entice desired employees to accept assignments there.

MOVING EXPENSES AND RELOCATION ALLOWANCES

Most firms pay the direct costs of moving employees and their families from one country to another. Companies are usually careful to specify what items they will pay to have moved or set weight limits on household items. Clearly, the most costly items associated with the move are household furnishings.

There are three major ways in which firms handle furnishing allowances. The first is that the company pays to ship the employee's furnishings to the new location. There is usually a maximum weight limit, such as 15,000 lb, on what may be shipped. The benefit to employees and their families is that they have their own furniture and home furnishings when they arrive at their new location.

Shipping furnishings can be expensive, and both damage and delays can lead to dissatisfied and sometimes angry employees. Consequently, many firms either purchase or more often lease household furnishings and provide them to international managers for free. Employees are often permitted to ship up to 1,000 lb of personal belongings. In addition, firms often pay for furniture to be stored during the assignment.

Storing furniture may seem like a simple task, but, if poorly done, employees can become quite dissatisfied, as this American was: "Almost all of our personal effects were ruined while in storage. It was difficult and extremely stressful to straighten the mess out. The company didn't really offer much help—we were on our own."

The third approach is simply to provide the employee with a fixed sum of money ($8,000 to $10,000) to purchase furnishings. If furnishings cost less than this amount, the employee is free to keep the difference. If they cost more, the employee must cough up the extra yen, pounds, rupiah, or lira.

In the event that the company pays a relocation allowance in addition to paying direct moving expenses, the relocation allowance is generally provided in recognition that a variety of expenses are incurred during an international move that cannot be accurately predicted and that vary by individual. These expenses are typically covered by a fixed allowance equal to 1 month's salary or $5,000, whichever is less. Between 30% and 40% of companies offer relocation allowances.

EDUCATION ALLOWANCES

Children's education is a critical issue in the minds of most parents who are asked to transfer to another country. The internal market at most firms is such that very few high-quality people would be willing to accept international assignments if their children were simply forced to attend local schools in the country where they were relocated. Consequently, most firms provide an education allowance that covers the normal costs (tuition, books, supplies, etc.) to attend local international schools. If adequate education facilities are not available in the country of assignment, many firms provide assistance that covers part of the cost to send a child to a boarding school in the country of origin. More comprehensive education allowances also include the cost of one or two round-trip tickets for the children to visit the parents. More limited allowances cover only airfare and not any of the costs to send a child to boarding school.

VACATION AND HOME-LEAVE ALLOWANCES

Approximately 40% of companies do not change vacation entitlement if an employee is on an international assignment. If the employee were entitled to 5 weeks of paid vacation prior to the assignment, he or she continues to receive this amount. How employees and their families spend their vacation time or the amount of money they do or do not spend is left entirely to the employees.

Some companies provide for home leave in addition to paid vacations. Companies that provide home leave do so for two primary reasons. First, they recognize that many families would stay on the assignment only if they could visit home regularly (usually once per year), and such a visit would not be required or the associated expenses incurred if they had stayed home. Second, most companies that provide this benefit want employees who are on global assignments to maintain ties and relationships with people in the home operations. They accept that this can be fully accomplished only with face-to-face visits and that without paid home leaves, these visits would be an extra expense for the employee.

Many employees and families we have interviewed strongly favor being given the equivalent sum in cash to use as they please rather than being required to take home leave. This allows them to have an extra paid vacation to go wherever they choose. Increasingly, firms either do not provide home leave and simply allow employees additional paid but not reimbursed vacation time or require home leave to be taken at home and reimburse expenses.

REST AND RELAXATION (R&R) ALLOWANCES

R&R allowances are most often associated with hardship assignments. Generally, they are provided so that the employee and family may get away and recover from the hardships of living in the country of assignment. These trips are often necessary to purchase goods or receive medical care not available in the country. Many firms have a "use it or lose it" policy with respect to R&R allowances; they do not want employees or their families to risk physical or emotional health for the sake of receiving the allowance money.

MEDICAL EXPENSES

The health of employees and their families is not something firms can afford to put at risk. Consequently, most firms pay for all medical expenses (often excluding optical and dental) for employees on international assignments. Some firms also pay for expenses in excess of what is covered by the employee's health insurance. In developing countries, this can often mean that firms pay for employees or members of their families to receive adequate medical care in countries other than the country of assignment.

CAR AND DRIVER ALLOWANCES

Except for senior executives, most firms provide a car allowance based on the differential between the cost to own and operate a car in the country of assignment and in the country of origin. External market pressures seem to be the biggest determinants of whether senior executives are given just a car or a car and driver. About one in four firms provide all employees from headquarters with company cars, whereas two out of three provide company cars when essential because of safety concerns or market demands. For example, in many Pacific Rim countries, American executives are provided with cars and drivers to which they would not be entitled in a similar-level position in the United States. Although there has been a reduction over the last 5 years in the number of companies providing this allowance, the change in policy has occurred slowly (over the last 5 years) because there are first-mover penalties for withdrawing perks from expatriates.

CLUB MEMBERSHIP ALLOWANCES

In many countries, club memberships are the only or the least expensive means for employees and families to gain access to recreational facilities such as tennis, swimming, exercise rooms, and so on. In other locations, club memberships are essential for gaining access to the informal but important places where business decisions and political and business contacts are made.

UNINTENDED CONSEQUENCES

Perhaps the most important unintended consequences of global assignments are those that occur not during the assignment but after. Although trying to enhance international mobility, some firms inadvertently create the opposite result. They entice people primarily with money and underleverage the power of growth and development and because of poor repatriation planning, career enhancement. This focus on money by the company creates a focus on money among employees. However, the fact is that many monetary incentives to go on international assignments disappear on returning home. Employees get frustrated by this, feel that there are no compensatory career payoffs, and as a consequence tell others to pass on any offers of international assignments the company tries to sell them. Consider the following comments from repatriated employees.

On returning, my base salary was reduced 16.7% versus the stated policy of a maximum of 10%. How is this possible when in the last 12 years I have received consistent outstanding performance reviews? These ratings have been by eight different managers, three of whom were senior vice presidents. In the last 3 years, I have received the maximum bonus available under the performance incentive plan.

—*American manager returning after three successive and successful assignments*

I was generally very unhappy after returning to the U.S. from Toronto, where the cost of living was less than in New York. As a result, even though I did get a raise and the additional benefit of a second lease car, I now have much less disposable income (higher taxes, higher housing costs) and, in fact, can't even afford one lease car because I can't afford parking for it! I am being hurt financially because I moved back to a much more expensive city and my company has no policy to handle that. It hurt particularly in contrast to the fact that I felt my overseas assignment benefitted me financially.

—*American manager returning from a 3-year assignment in Canada*

As is clear from all the allowances most firms provide their international employees, maintaining these employees can be expensive. One multinational U.S. firm with 500 employees on international assignments estimated that the incremental cost paid directly and indirectly to these employees was $80 million. As mentioned, some firms try to reduce the total costs of maintaining global employees by slashing the number of employees it has on assignment. The fastest way for a firm to cut $40 million from its $80 million total is to cut the number of expatriates from 500 to 250, as one West Coast bank did recently. However, this move, which was intended to make the bank lean and mean, actually left it anorexic.

LEAPING AHEAD TO THE PAST

No one would argue that maintaining employees abroad is not expensive. In addition, in their efforts to make their international assignment compensation systems equitable, many have become quite complex. In combination with the expense, this may explain why more firms would like simply to set a percentage premium and leave it at that. Increasingly, policy makers are saying, "Give these managers an extra 20% and send them." Interestingly, these changes in expatriate compensation actually represent a return to the past. In the early days of sending employees abroad, say in the 1950s, most managers were given a little extra and off they went. However emotionally appealing this simplification might be to frustrated human resource directors and other executives, it is likely to be disappointing because of the problems it doesn't solve. For example, it won't solve the problem of how to attract high-quality people to accept international assignments. It won't solve the problem of the significant effect exchange rate fluctuations can have on real purchasing power in another country. It won't solve the problem of lowered standards of living due to high inflation rates in the country of assignment. It won't solve inequities employees feel when they compare themselves to employees at similar levels in other firms.

Compensation systems are expensive and complex. However, the focus should be on the entire reward system and then on the effectiveness of the compensation systems. From a strategic perspective, simply cutting the number of employees on international assignments may do nothing positive and may actually have negative effects on organizational and individual effectiveness. Likewise, merely simplifying compensation systems may increase efficiency at the expense of effectiveness. Many people and systems are expensive and complex. That is not the issue. The real issue is how to make sure that costly and complex assets are worth it.

A BROADER PERSPECTIVE on Comp.

One way to take a broader view of the issue of compensation is to answer a seemingly narrow question: "Why do I have to pay so much extra to get people to accept overseas assignments?" The simplified answer is that human motivation is such that to induce someone to do something, anticipated benefits must be greater than anticipated costs. To the extent that employees perceive the firm as taking an out-of-sight, out-of-mind approach to international assignments, the career costs can be substantial. In our research, most managers (regardless of nationality) felt that their firms did not value international experience (Black & Gregersen, 1991; Gregersen et al., 1995; Stroh, Gregersen, & Black, 2000). Facts such as these and horror stories about so-and-so being caught in a holding pattern for 6 months after returning from overseas only serve to raise the career costs in people's minds. Some people are willing to disregard these career costs for extra compensation. The more talented an individual thinks he or she is and the greater uncertainty of a career payoff after the international assignment, the greater the potential costs to that individual and consequently the greater the compensation the firm must pay to induce the individual to accept the assignment. It's simple economics and math.

There are, however, many positive aspects or benefits to accepting an international assignment independent of the monetary compensation. Most employees expect—and actually experience—greater job autonomy and responsibility while abroad. They develop market knowledge, language skills, contacts, and global GM perspectives. Certainly these are positive factors in the cost–benefit analysis. For most employees, they would be. However, it seems unlikely that these short-term benefits will simply cancel out the anticipated long-term career costs, especially if the firm is unlikely to utilize these skills and experiences in the future. It takes the employee time and energy to acquire the language skills and market knowledge. To the extent that the future career payoff is uncertain, economics argues that there must be certain monetary payoffs today.

The point of all this is not to try to examine all the potential costs or benefits that might go into employees' decisions to accept or reject international assignments; rather, the point is that both monetary and nonmonetary considerations go into making their decisions. This is why simply focusing on monetary compensation is a formula for rewarding A (maximizing short-term monetary benefits over the costs to the individual) while hoping for B (focusing on both long-term monetary and nonmonetary factors being taken into account in the decision to accept the assignment and in performance).

One simple means of reducing but not necessarily eliminating the escalating costs per individual manager sent abroad is to increase the rewards managers are given for successfully completing the assignment. In the chapter on repatriation, we talk specifically about how to redesign repatriation policies so as to maximize their positive influence on employees' perceptions of the potential rewards of working abroad. We argue that at least some of the extra monetary rewards could be replaced by nonmonetary rewards such as better policies for employees sent abroad, career systems, and repatriation procedures. Simplified, if employees thought that the policies for international employees would enhance their ability to perform well during the assignment, and if they believed that their performance abroad would be rewarded on repatriation, they would be motivated to accept these assignments even if there were fewer immediate monetary inducements. Notice that we do not say that they would be willing to forgo all monetary inducements. These ideas can be illustrated in the framework of a well-known theory of motivation—expectancy theory (see Fig. 8.2).

First, employees ask, "What is it I am expected to do?" In other words, is the performance target clear? If it is not, it is like shooting in the dark. Where is the motivation in that? Although this may seem obvious and self-evident, in one survey, only 33% of firms said they always set measurable goals for employees on global assignments, and 13% said they never set measurable goals (Halcrow, 1999). A total of 67% of the companies indicated that performance on international assignments had been hurt by the lack of defined goals or targets.

Next, employees ask, "If I try, can I hit the target?" Appropriate selection, training, and cross-cultural support practices and policies can all enhance the percentage of positive answers to this question. Although firms might think that any employee they have worth his or her salt would have to answer "yes," it is important to keep in mind that employees do not ask or answer this question in the abstract. They ask and answer it in the context of what they know (i.e., how they have done at home) and what they know less about (i.e., how they would do in the international assignment). Thus, not providing the required training and support can seriously undermine or even destroy the anticipated and desired motivating force of the compensation package.

Finally, employees ask, "If I do it, what will I get?" Here the answer is a function of math that we all learned in high school—expected value. The motivating impact of an outcome is not just a function of how much a person values the out-

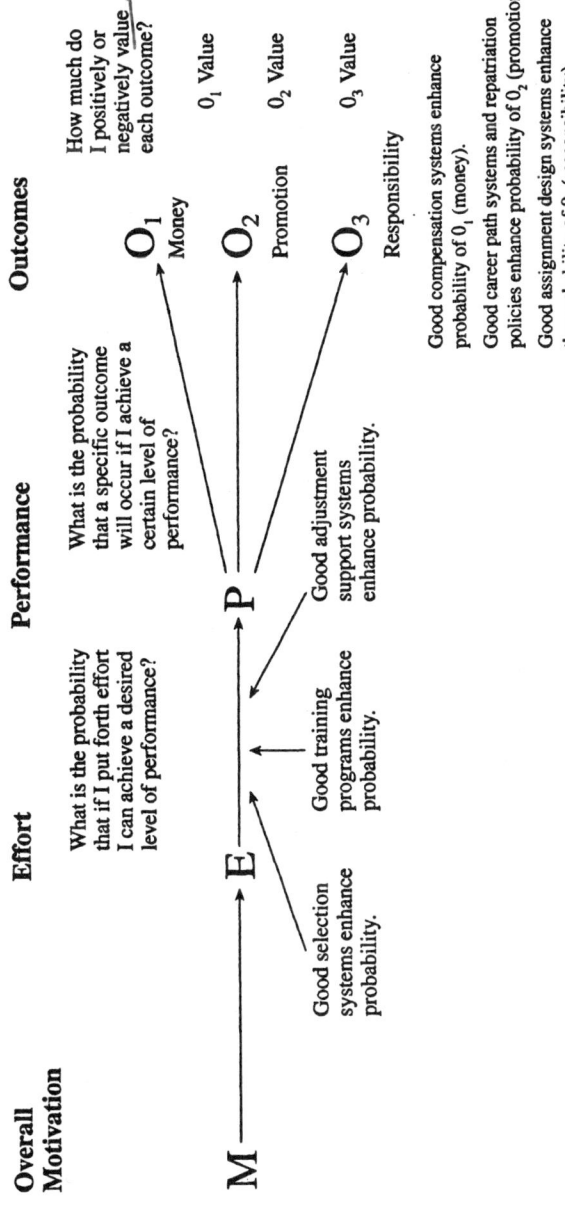

FIG. 8.2. Expectancy theory and global assignments.

come but the value of the outcome times the likelihood of outcome (i.e., the expected value). For example, an employee may place a high value on the long-term career payoff of an international assignment (let's weigh it 50 out of a total 100 points), but if the individual believes that the likelihood of that payoff is only 10%, then the expected value is only 5 out of 100 points.

Here it is important to recognize that perception is reality. It doesn't matter what senior executives think or what international compensation managers believe are the payoffs or benefits of an international assignment. All that matters is what the individual employee considering the opportunity believes. Thus, the first thing to make sure is well understood is that the employee believes there will be payoffs for working abroad. Even if the employee's perceptions are wrong, unless the firm knows what they are, it can't begin to try to reduce the perception–reality gap either by reshaping mistaken perceptions or changing dysfunctional company practices and policies.

CONCLUSION

Most managers responsible for international assignment policies and practices have told us that compensation gets the lion's share of time and attention. It's easy to understand why—taxes are complicated, and computing housing, goods and services, and other cost-of-living comparisons is expensive and time consuming. Instituting an inappropriate compensation policy can easily mean failure to meet any or all three of the primary objectives of mobility, equity, and cost control. However, compensation alone does not provide the full picture of employee motivation.

Sending people on international assignments is costly and the reward systems are complex. The critical issue, however, is the cost effectiveness of the assignments and the associated reward systems. Unfortunately, the overemphasis on monetary rewards has been in part responsible for the problem of rewarding A while hoping for B—of maximizing monetary incentives while hoping for a recognition of other, nonmonetary rewards. We have argued that good selection, training, adjustment, and repatriation policies can not only enhance the probability of effective performance during and after an international assignment but can also be effective in shifting the focus to the nonmonetary rewards and increasing an individual's motivation to accept and do well in an assignment. Consequently, a good international assignment compensation system must include both effective monetary and nonmonetary rewards. Although many firms want to reduce the total cost of maintaining overseas operations, we have argued that firms should first be concerned with the cost effectiveness of individual global managers. By focusing on cost effectiveness, a company reduces the likelihood of cutting costs while hurting mobility and equity objectives.

Our contention is that for international assignment compensation systems to maximize all three of their objectives, they need to be placed in a larger framework of employee motivation. The simplest and most comprehensive framework we have found is expectancy theory. Money and its representation in allowances, differentials, and the like is only part of the outcome picture, and outcomes are only part of the overall employee motivation picture. It is only when placed in the proper overall perspective that each piece of the picture can be properly viewed, evaluated, and shaped.

PART
IV

AFTER THE ASSIGNMENT

CHAPTER

9

Repatriating: Helping People Readjust and Perform

For many employees and their families, repatriation conjures up images of joyous reunions and settling comfortably into a familiar life back home. All too frequently, though, the reality is very different; most repatriates are lucky to receive any welcome at all. As the spouse of one American expatriate put it, "If you look at repatriation as a homecoming, you're setting yourself up for failure." Sadly, in many cases, this sense of failure encompasses every facet of the returning employee's life.

Why is adjusting to life back home so difficult for most employees and their families? What can companies do to better prepare employees for repatriation? Until recently, these questions received little attention from either researchers or global companies. This is beginning to change as organizations seek answers to why rates of job dissatisfaction and turnover are consistently higher among managers returning from global assignments than among managers who remain at home. According to one study (Black & Gregersen, 1999), 25% of all managers who complete their global assignments leave their companies within a year after returning home. (The figure for managers who remained at home was half this.) In some European and U.S. companies, the turnover rate is as high as 55% within 3 years after repatriation.

CHALLENGES OF REPATRIATION

Most employees beginning global assignments assume that they and their families will go through a period of adjustment as they get used to the sights, smells, foods, and customs in their new culture. They may even expect to experience the symptoms of culture shock discussed in chapter 2 (this volume). However, most employees never expect that adjusting to being home will be anywhere near as difficult as the process they experienced when they first relocated. In fact, as the following comments reveal, repatriation can be as difficult, if not more difficult, than adjusting to a new and unfamiliar culture (Rodrigues, 1996).

> Coming back home was more difficult than going abroad because you expect changes when going overseas. It was real culture shock during repatriation. I was an alien in my home country. My own attitudes had changed, so that it was difficult to understand my own old customs. Old friends had moved, had children, or just vanished. Others were interested in our experiences, but only sort of. They simply couldn't understand our experiences overseas, or they just envied our way of life.
> —*Expatriate spouse returning from 3-year assignment in Vietnam*

> Treat coming home as a "foreign assignment" and spend time getting the "lay of the land." Don't expect any special treatment—you're basically a "new hire." Look out for yourself, because no one else will. After being home for nine months, and after giving 33 years of my life to this company, I still have no office to work in—just a "bullpen" with a temporary assignment.
> —*Expatriate returning from 1-year assignment in England*

> Now that I'm home, it seems like my overseas assignment is a punishment, a real "ball and chain," in terms of my career.
> —*Expatriate with 14 years of international experience*

To some top executives and line managers (especially those without international experience), these comments may seem a bit overstated, but our research on repatriation found that 60% of American, 80% of Japanese, and 71% of Finnish expatriates experienced some degree of culture shock during repatriation (see Fig. 9.1). Our results are similar to those Adler (1981, 2000) has found in her study of American repatriates—namely, that the culture shock of coming home is usually more difficult than the culture shock of relocating for a global assignment (Black, 1994).

As the reader shall see, the process of cross-cultural adjustment for returning managers and their families is very similar to the process of cross-cultural adjustment discussed in chapters 2 and 5 (this volume). In short, people often return home from global assignments with incorrect mental maps of what to do, learned inabilities regarding how to do it, and uncertainties about what the results of their actions will be in the now-"foreign" home country.

Unfortunately, most executives have little sympathy for the problems of repatriating employees and their families (Black & Gregersen, 1999). In fact, the home-office response often goes like this: "Culture shock coming home? What's the big deal? After all, they're home."

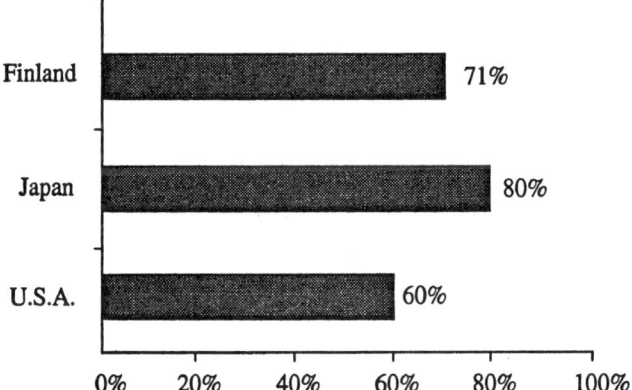

FIG. 9.1. Repatriation culture shock. Note: This chart shows the percentage of managers who found repatriation adjustment more difficult than the original adjustment overseas.

Contrary to what most managers in the home country think, repatriates have to adjust to significant changes when they return home. These changes may include new political systems, transportation systems, social groups, eating habits, and so on—in short, many of the same components of the culture that were unfamiliar when the employee first moved abroad. Of equally great importance, the employee is not the same after an international assignment of 3 to 5 years (Stroh, Gregersen, et al., 2000).

THE REPATRIATION PROCESS

Without a doubt, employees and their home countries change throughout an international assignment. Some of the major components of these changes are outlined in Fig. 9.2. Before the assignment, employees consciously and unconsciously acquired mental maps and behavioral routines that worked effectively for them in the home country. While living in another country for an average of 2 to 5 years, most of these employees acquired new mental maps and behavioral routines. As the employee adjusted and changed in the new culture, objective aspects of the home country were changing simultaneously. For example, consider the sweeping political and social changes that have occurred since 9/11.

Besides potentially dramatic changes, endless little changes occur in the home country while the employee is on the global assignment. These changes can occur in a host of contexts including in neighborhoods (people move in and out), friendships (old friends get new friends), and schools (funding levels may shift or teachers change). In combination, all of these changes contribute to difficulties adjusting, to the overall level of satisfaction or displeasure with being home, and

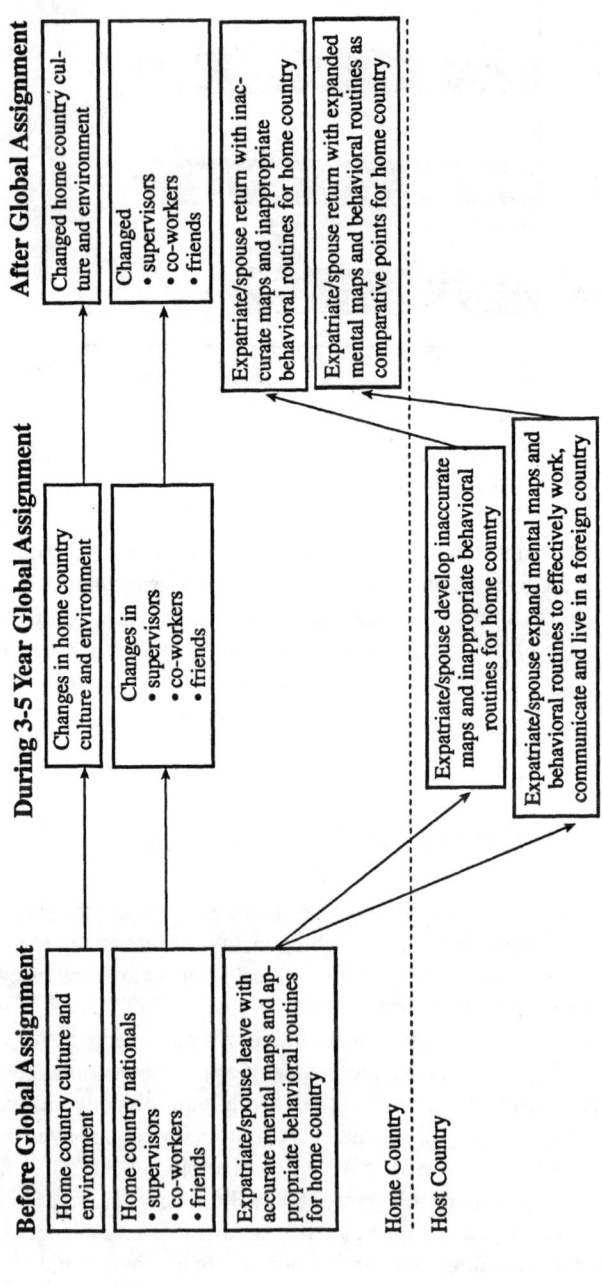

FIG. 9.2. Components of change.

accordingly, with the high turnover rate among returning managers (Stroh et al., 1998).

Dramatic as well as small changes can also occur in the firm (e.g., restructuring, strategic shifts, managerial advancements, etc.). Some of the changes and their potential impact on adjustment are reflected in this manager's experience after working in the parent company for more than 18 years:

> The division I worked for was reorganized, and the subsidiary I worked for was placed under stringent cost-cutting guidelines. My reentry was very cold, with little support in finding a job since previous management had been fired. Instead of placement, I had strong feelings of displacement when coming home.

Further complicating the situation, in addition to objective changes in both the home country and the parent company, subjective perceptions of what used to be and what should still be also change throughout a global assignment. These changes in mental maps occur for several reasons.

First, individuals have difficulty remembering exactly what used to be, and if they do remember, these memories often tend to focus on the ideal aspects of home. This process could be called the *Dorothy Syndrome* after Dorothy in the *Wizard of Oz* who felt there was "no place like home" as she fondly recalled her life in Kansas. Reflecting on this perceptual distortion, one spouse commented

> All of the years that I lived overseas (largely in underdeveloped countries), I always thought that the U.S. was really better, more efficient, and so on, than anywhere else in the world. For 26 years and four global assignments, I carried this idea around: Now that I am finally back home, I am finding out that things in general are just as inefficient in the U.S. as in other countries. This has been a very difficult reality to accept.

When expatriates and families return home after several years abroad, their supervisors, coworkers, and friends have also changed. However, these various people often incorrectly assume that the returning employee and his or her family are the same as they were before their global assignment. One spouse returning to Finland after 6 years abroad described this dynamic well: "Family and friends didn't want to admit that I had 'grown up' during the years in a foreign country." An American manager expressed similar sentiments: "Previous friends expected me to return unchanged and resume life as if I had not left."

REPATRIATION ADJUSTMENT AND THE BOTTOM LINE: PERFORMANCE

In our work with multinational firms around the world, executives frequently ask, "Why should multinational firms pay attention to the adjustment of expatriates and their families during repatriation?" Our response, based on research and experience, is simple: Failure to pay attention can have a negative impact on the bottom line. In other words, companies that fail to prepare employees for repatriation are likely to experience higher turnover rates among returning managers as well as reduced performance among executives and managers who stay on. To put these findings more positively, when repatriation goes smoothly, managers are more likely to perform better in their jobs (Stroh, Gregersen, et al., 2000).

Furthermore, when the family of an employee adjusts during repatriation, a positive "spillover" effect occurs in which the positive home situation spills over to work, resulting in better performance by the manager (Stroh, Gregersen, et al., 2000). The reverse is true as well: When a partner has difficulty adjusting, it is likely to be reflected in the performance of the employee. An American repatriate we interviewed emphasized this link: "My spouse has had a very difficult time coming home from Europe and living in the suburbs of America. She hates it. Her adjustment difficulty has made my life less than wonderful and my work performance less than excellent."

DIMENSIONS OF ADJUSTMENT UPON REPATRIATION

We have identified three basic areas that managers and their families adjust to when returning home; they parallel elements of the process of cross-cultural adjustment international assignees and their families experience following relocation (Black, Gregersen, & Mendenhall, 1992). First, repatriates need to adjust to new jobs and work environments. Even though most repatriates have almost 15 years of experience in a parent company, they still make comments such as, "Be prepared for corporate culture shock when you come home!" One manager with 12 years of experience in the parent company commented, "Our organizational culture was turned upside down. We now have a different strategic focus, different 'tools' to get the job done, and different buzz words to make it happen. I had to learn a whole new corporate 'language.' " Interestingly, work-related adjustment challenges during repatriation were one of the most frequent problems mentioned by the managers we interviewed.

Second, expatriates and their families need to adjust to communicating with home-country coworkers and friends. After a global assignment, people at home often seem like foreigners. For example, Americans are well known for making small talk at the beginning of a conversation, whereas people from other cultures may find this mode of conversation insincere. After a Finn spends several years in the United States and learns small talk, he or she returns home to fellow Finns who react quite negatively to such "trivial" conversations.

Another common challenge of communicating with people at home is that they are not generally interested in the employee's experiences while living abroad. Finally, children often encounter significant language difficulties during repatriation. This situation is especially true for younger children born during global assignments who often fail to learn the complexities of their home-country language and for teenagers who are very aware and self-conscious of differences between them and their friends in such matters as accents, knowledge of slang, and intonation. Communicating effectively with people during repatriation is a challenge, as reflected in the following comment:

> After coming home, my daughter felt neither British nor American from a cultural standpoint. She went from being president of the school overseas to being a new face at home. Nine months after our return, she still feels quite "different" from the other students.
> —*Expatriate returning from fourth global assignment in 9 years*

Third, expatriates and their families face the problem of readjusting to the general living environment (e.g., food, weather, housing, transportation, schools, etc.), even though they usually lived in their home country for most of their lives. These adjustments to the general culture are often the most challenging. The following comments describe some specific dilemmas:

> It was challenging to return home and find housing, locate stores, and make friends. Even though I'd lived in the same metropolitan area and country before, this was like moving into a new world, and I had to start from scratch. I never realized that in returning home I would not be instantly home.
> —*American spouse after 6 years in Europe*

> I was totally unprepared for the long, harsh, cold winters ... even though I had grown up in Finland.
> —*Finnish expatriate after 3 years in Australia*

> I never realized how difficult and exhausting simply commuting to and from work is in Japan until now. I hate it.
> —*Japanese expatriate after 5 years in the United States*

FACTORS INFLUENCING REPATRIATION

Research on the three aspects of adjustment just discussed has found that certain factors inhibit one or more of these aspects of adjustment. Figure 9.3 summarizes those factors that affect adjustment before individuals return home and those factors that that influence adjustment after they return home. These factors are further categorized into several groups including important sources of information about changes in the home country and the parent company and individual, job, organization, and nonwork factors that affect adjustment after returning home.

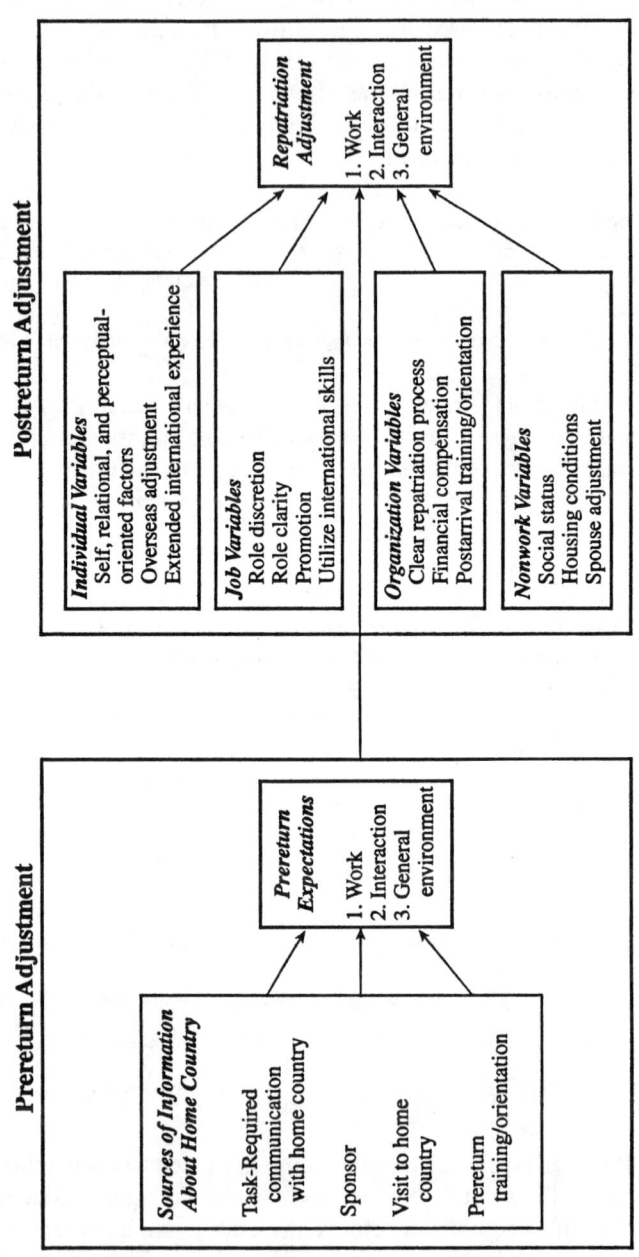

FIG. 9.3. Balance sheet.

We discuss each of these factors and how they influence the adjustment of managers and their families during repatriation.

MECHANISMS TO FACILITATE ANTICIPATORY ADJUSTMENT

Just as people make anticipatory adjustments before embarking on a global assignment (see chap. 5, this volume), individuals also make adjustments before transferring home. These adjustments are primarily mental in nature. In other words, people begin to make changes in their mental maps of what work and living will be like even before they actually return home. For repatriation to be successful and for the returning employee to want to stay with his or her employer, the repatriate's expectations must be realistic (Stroh et al., 1998). Several potential sources of accurate information about the home country can help modify and mold repatriates' expectations and mental maps.

Information Exchange

Managerial and executive jobs often require extensive interaction between company headquarters (in the home country) and a foreign operation. Job- or task-required interaction and information exchange is especially necessary in multinational and global firms, which typically require a great deal of coordination between headquarters and foreign operations. Because of these coordination requirements, a reasonable amount of accurate information is likely to be passed on. It is important to remember, however, that most of this information is undoubtedly focused on changes in the parent company; relatively little is related to changes in neighborhoods, children's schools, and friendships outside work. Accordingly, information acquired through job-required interaction is most likely to facilitate an employee's adjustment to work but will have less impact on the employee's adjustment to nonwork-related aspects of the employee's life.

Sponsors

Another source of primarily work-related information is an organization sponsor, mentor, or "godparent." A formally or informally assigned mentor can provide an employee on a global assignment with important information about structural changes, strategic shifts, political coups at work, promotion opportunities, and general job- and company-related knowledge. This information is not likely to help people adjust to the general culture; it is more likely to help them effectively adjust to communicating with people at home. Unless an employee has an effective sponsor during and after a global assignment, the employee may end up facing any number of dilemmas. One repatriate recounted his experience. "After spending 9 years in Venezuela, fitting in is difficult if you have no ongoing contacts. There is tremendous insecurity, since no knows you or cares, and you can easily get caught in the next job-reduction plan. Some reward for all the sacrifice of going abroad!"

In this case, an effective sponsor might have reduced some of the problems this employee encountered. Even though sponsors can provide important adjustment-related information during global assignments and helpful support during repatriation, relatively few individuals have designated sponsors throughout the time they are abroad. In fact, our research found that only 22% of American, 22% of Japanese, and 51% of Finnish expatriates had sponsors (Black, 1991). This is unfortunate because individuals with sponsors generally adjusted better at work than those without them. More generally, maintaining contact with employees who are out of the country is an effective way to help prepare returning employees for repatriation while providing them with information they will need to know (Paik, Segaud, & Malinowski, 2002).

Research suggests that women managers may benefit even more from having a sponsor or mentor than male managers both during and after returning from assignments abroad. In interviews, women have noted that their mentors eased their reentry at work and ensured the women were not out of sight and out of mind (Linehan & Scullion, 2002).

Home Leave

Making periodic visits home is another effective way for employees on international assignments to stay in touch with coworkers, friends, and family members and thereby keep abreast of changes at work, in their social circle, and in the home country. Visits also give coworkers, friends, and family members a chance to observe changes in the person who has been living abroad. Whereas 65% of the managers we studied received paid home leaves for themselves and their families throughout their assignments, a study of approximately 250 American and Canadian multinational firms found that only 35% of managers and families are required to take these leaves at home, preferring instead to take the family to more exotic places (Black et al., 1992).

Orientation

In earlier chapters, we discussed cross-cultural training. As we mentioned, research suggests that such training should be provided both before and after relocation (see chap. 4, this volume). Ideally, training should also be provided during repatriation. Repatriates of various nationalities have emphasized the importance of offering training or an orientation after employees return home. In fact, except for help in locating appropriate assignments, repatriates rank training highest among benefits they would like multinational firms to provide after employees return from abroad (Black, 1991). Training during repatriation can facilitate adjustment not only at work but also in communicating and living in the home country.

Given the research support for prereturn training, it is somewhat surprising that relatively few multinational firms provide such training or orientation. According to data we collected, only 36% of American, 8% of Japanese, and 23% of Finnish employees receive any training before or after repatriation. Further, only about 10% of all spouses/partners receive such training (Gregersen & Black, 1992). One spouse suggested that "a repatriation meeting should be provided to families [after they return home], like the orientation meeting before going overseas, since many changes occur at home in 3 to 5 years." Without training, many expatriates and spouses inefficiently search for (and often do not find) accurate information about their home country after they return.

POSTRETURN ADJUSTMENT

Several factors can facilitate or inhibit repatriates' adjustment to work, to communicating with people in their community, and to the general culture. These factors can be grouped into individual, job, organizational, and nonwork categories.

Individual Factors

Many of the important individual factors relevant to effective cross-cultural adjustment were outlined in chapters 3 and 5 (this volume). These factors apply equally well to the process of repatriation. Specifically, self-oriented factors, relational factors, and perceptual factors can have a positive impact on all facets of the adjustment process.

Ironically, successful adjustment overseas can result in more significant challenges on returning home. This scenario is especially true if the culture to which the employee and family relocated is very different from the home country and the assignment has been a long one. Essentially, the more people acquire the maps and rules of the country where they were relocated, the more difficult reacquiring the maps and rules in the home country will be. A comment by an American who was abroad for 21 years sums up the problems of adjusting after repatriation (although most managers are not gone for 21 years): "Twenty-one years overseas was terribly hard to 'shake' after coming home." Other people who were on multiple global assignments or especially long assignments noted that they felt like "aliens" in their native lands.

Job Factors

One of the most important components of a successful adjustment to being home is the job assignment or position the employee is given. Unfortunately, according to employees surveyed for one study (Halcrow, 1999), 25% of U.S. firms do not begin formal discussions about the next job assignment until 3 to 6 months before the employee's global assignment is expected to end; and 27% say that their companies do not necessarily have any discussions about repatriation at all.

Further, employees on global assignments often receive little or no notice that they will be repatriating. According to one survey (Halcrow, 1999), as many as 18% may get no formal notification. That firms give employees such little notice that they will be going home epitomizes the lack of concern given to repatriation among U.S. companies. This lack of concern is further reflected in the negative comments recently repatriated employees make about their job experiences. The following comments are typical:

> After being home 3 months, I am still waiting for a permanent office. All this after 33 years of experience in the company and three international assignments!
> —*American expatriate*

> My job description did not even exist when I came home. I felt as though I had no status in the company. In fact, everybody was saying, "Hey, what are you doing here?" It seemed like I was just a temporary "extra" for at least nine months.
> —*Expatriate with 9 years of experience in the parent company*

> No one accepted responsibility for placing me back in the organization. I ended up without a job, when I was expecting a promotion! My wife also gave up her job with the same company to go overseas (she had 12 years of experience). We were promised a job for her upon our return. Again, no one has helped us find one.
> —*American expatriate with 15 years of experience in the parent company*

Ill-Planned Job. From these accounts, it is clear that the first step in effective adjustment at work is having a job. Nevertheless, approximately two thirds of expatriates, regardless of nationality, do not even know what their assignments will be before they return (Black, 1991). Because return assignments are so often unplanned, repatriated employees often refer to "holding pattern" return assignments, which resemble airplane holding patterns over congested airports. It should come as no surprise that when employees returning from global assignments did get jobs, these positions were rarely optimal. In fact, they were frequently ill-defined, low-impact, "make-work" positions intended to keep the employees occupied and out of the way.

Reduced Job Autonomy. When permanent positions are located, they often have reduced autonomy and authority compared to the position the employee held during the most recent overseas assignment. Research indicates that nearly half of all repatriated managers have less autonomy and authority after returning home (Black, 1991). The following comments are typical:

When I was overseas, I felt I had an impact on the business. In the U.S., I felt as though the impact—if any—was minimal. When I came home, I was assigned to a newly created, undefined staff job where I had no friends, no contacts, and no access to management. Firms need to realize that expatriates have developed independent decision-making skills, have become accustomed to having final authority, and are conditioned to having their business judgment given a lot of credibility by top management. In my new job, my business judgment is much less valued than when I was overseas. Until firms change, expatriates should expect the worst when coming home to avoid disappointment.

The importance of providing repatriates with high-discretion jobs cannot be overemphasized. Disappointment, frustration, and often a desire to look for greener pastures are just a few of the reactions repatriates are likely to experience if they are given jobs with less authority than they had overseas.

Unclear Job Expectations. In addition to job discretion, repatriates should be given positions with clear job descriptions or high role clarity. This also facilitates adjustment at work. Undoubtedly, the first step in clarifying jobs for repatriates is finding them something better than holding pattern positions. Temporary positions are guaranteed to be highly ambiguous because they will essentially be make-work assignments.

Promotion Disappointments. Many employees take global assignments in the hope of receiving a promotion after successfully working overseas. Unfortunately, this is a romanticized myth that does not usually match the reality of repatriation. As one European manager with 20 years of international experience in eight assignments said, "When you go overseas, your firm absolutely forgets you in terms of promotions. In fact, my overseas assignment seems more like a punishment, in terms of my career." For Americans, the picture is not much brighter: "I went on my foreign assignment to the U.K. as a favor for the department. In return, I received nothing special for the 10 months I spent away from my family and the hardship I put them through. I really expected a promotion after coming back home and did not receive it." In reality, only about 10% of American and Japanese managers receive promotions over the positions they held overseas; the rate for Finnish managers is almost 30% (Black, 1991).

Demotion Surprises. These low rates of promotion may reflect the significant restructuring and downsizing occurring in many multinational organizations. It is unlikely, however, that the high percentage of demotions was matched in the managerial cohort at home. Specifically, on repatriation, 77% of American, 43% of Japanese, and 54% of Finnish managers were demoted to lower level positions than they held overseas. Equally telling, the majority of employees who have gone on international assignments feel that their global experience had a negative impact on their careers (Adler, 2001; Hammer, Hart, & Rogan, 1998).

Poor Skill Utilization. Expatriates often gain unique country knowledge, language proficiency, and international management skills during global assignments. In fact, one of the strategic purposes of overseas assignments is to develop these skills and knowledge. After employees return, however, their skills are utilized inconsistently. Less than half of the returning managers we surveyed reported that they were utilizing their international experience since they were home. After making million-dollar investments to send, support, and bring home each expatriate, it is surprising that firms are willing to obtain such minimal returns on so many of those investments. Returning managers have similar feelings, as these comments suggest:

> Firms must value international expertise . . . not only appreciate it but actually put it to good use. Don't let a corporate headquarters environment destroy the lessons, "business savvy," negotiation skills, and foreign-language proficiencies that expatriates learned from the real world . . . a global marketplace.
> —*American expatriate*

> This company places little value on my international skills. In fact, I am now fluent in Japanese, but the company has shown no interest in placing me in a position to use my recently acquired language skills. What a waste!
> —*American expatriate*

> I hope I will be able to use my knowledge about Europe in my company in the future, but now I don't feel as if this knowledge is appreciated.
> —*Japanese expatriate*

Organizational Factors

The parent company's overall approach to the repatriation process and whether employees think the company provides adequate financial compensation during the assignment can have a significant impact on adjustment after the employee returns home. Unfortunately, only a minority of firms systematically manage the repatriation process.

Unclear Expectations. Most returning managers feel that their companies gave them a very unclear picture of the repatriation process. Many were uncertain and concerned about the nature of their positions when they returned home, career progression, compensation equity, taxation assistance, and so on. This sense of uncertainty is reflected in this American's comment: "What was it like coming home? No clear direction in my job, no clear direction in my career, and no assistance during the return. You are on your own around here!" Undoubtedly, the overall ambiguity surrounding the repatriation process is largely a reflection of the nonstrategic approach firms take to global assignments in general.

Financial Shock. In addition to providing a clearer picture of the process of repatriation, firms need to pay much better attention to the financial compensation packages they offer managers when they return home after a stint abroad. Because nonmonetary rewards received during repatriation are generally quite low (temporary job, demotion, etc.), repatriates pay particular attention to monetary rewards and to potential shifts in living standards. Some of these financial dynamics are captured in the following remarks:

> The cost of everything—from housing to the basic necessities—was so much higher in New York that it was literally shocking, even though we expected it. Everything is so expensive in New York City. In Mexico and, before that, Brazil, my family and I lived quite well. After returning to the states, it was back to reality—purchases had to be budgeted, something we hadn't done in years.
> —*American expatriate*

> When you are overseas, you receive many benefits, such as maybe a free car, or free gas, or a nicer home than you had back in your home country. You become accustomed to these benefits. Then when you return to your home country, things return to normal, and lots of these benefits you were accustomed to disappear. It's like being Cinderella, and midnight has struck.
> —*American spouse*

For these two families, the balance sheet for expatriate compensation (see chap. 8, this volume) is clearly out of balance. Apparently, they received many of the allowances available to expatriates during the assignment, but these allowances are generally unavailable after the employee comes home. Much like drug addicts, repatriates and their families often experience financial "withdrawal" as they struggle to adjust to their new standard of living. Roughly three fourths of all expatriates, regardless of nationality, have to cope with significant decreases in their standard of living after returning home (Black & Gregersen, 1999).

No Repatriation Training. Because many international firms do not pay systematic attention to repatriation compensation and fail to clarify the repatriation process, it should come as no surprise that repatriates receive very little training and orientation after coming home. The overall lack of company-provided training and orientation is well reflected in this American spouse's comment: "Give more information about the home country, and about all the things the company is willing to help with-no matter how small. Only one or two of the wives I know received any formal information about the company, Most of what we found out had to be dug up and passed on by word of mouth, from one to another. As a consequence, much of the information we received was too late to do most of us any good." Providing training and orientation after a global assignment is an effective way to enhance adjustment.

Nonwork Factors

Two primary nonwork factors influence adjustment after repatriation: shifts in social status and changes in housing conditions.

Drop in Social Status. After coming home, employees and their families lose the formal status of being foreigners. In one study (Black & Gregersen, 1991), 54% of the American, 47% of the Japanese, and 27% of the Finnish repatriates indicated that they experienced a significant drop in social status after they returned home, whereas fewer than 4% of the total group of respondents said their social status increased. As reflected in these statistics, repatriates and their partners are less likely to be treated as guests of honor at social and recreational functions (dinners, receptions, etc.) or as "guests" in their neighborhoods. As one spouse said, "The biggest surprise of coming home was that I simply didn't realize how specially we were being treated in all aspects of life during our international assignment."

During an overseas assignment, expatriates typically feel like big fish in a little pond. After coming home, they are little fish in a big pond. The demotion, loss of financial perquisites, and absorption into corporate headquarters that typically accompany repatriation can easily increase the sense of lowered social status. According to one American returning from a 9-year executive assignment in the United Kingdom, "If you have been the orchestra conductor overseas, it is very difficult to accept a position as second fiddle when coming home." In general, net losses in social status have a negative impact on all facets of repatriation adjustment.

Housing Conditions. In addition to shifts in social status, changes in housing conditions can significantly influence adjustment among repatriates. Specifically, whether they have appropriate housing when they return home is positively related to all three facets of repatriates' adjustment and to the general adjustment of spouses and partners during repatriation.

Three major issues appear to influence perceptions about the quality of the housing. The first issue is whether and for how long the family has to stay in a hotel before moving into permanent accommodations. Families that rent their houses in their home country while they are abroad usually have to stay in a hotel for 2 to 15 weeks while damage caused by renters is repaired. These costs can run to more than $15,000 (after only 2-year assignments), and only rarely do firms compensate repatriates for these losses.

The second issue is how quickly the employee has to find housing. Repatriates who sell their homes before they relocate often have very little time to find housing after they repatriate. In some cases, hotel allowances are cut off before housing loans can be approved. In other cases, employees feel forced to make less-than-optimal housing purchases because of the hassle of living in a hotel.

The third issue is whether finding suitable housing is a problem. Finding appropriate housing can be a major challenge if the repatriate sold his or her home before the global assignment and housing prices rose while the employee was abroad. More than half of all North American multinationals do not give repatriates housing allowances. In a worst-case scenario, the employee may be forced to buy a less desirable home than the employee and his or her family lived in before they went on the global assignment.

Our research found that Japanese repatriates were the most likely to experience a significant decline in the standard of their housing. In fact, almost 70% of Japanese repatriates we surveyed stated that their housing conditions during repatriation were less than satisfactory (Black, 1991). (The Japanese face a somewhat unusual challenge in that Japanese houses and apartments are generally much smaller than the accommodations they have overseas.)

UNIQUE ADJUSTMENT CHALLENGES FOR SPOUSES

Many of the predeparture and postarrival factors that we have discussed so far are relevant to both employees and their partners or spouses. However, spouses and partners sometimes experience unique difficulties.

Career Challenges

In our study on repatriation, we found that 55% of American spouses worked before, 12% worked during, and 30% worked after their global assignments. Seventy-two percent of the Finnish spouses worked before, 20% worked during, and 75% worked after their assignments (Black, 1991). These data suggest that many women, especially, make significant career sacrifices so that they may accompany their husbands abroad; however, many seek employment after they return.

> The biggest challenge of coming home was going back to work again. Years without training and schooling resulted in a big career loss ... not to mention the pension loss.
> —*Finnish spouse returning from 3-year assignment in Saudi Arabia*

> My contacts and visits with work colleagues in the home country were very occasional during the international assignment, and the reestablishment of those contacts and my return to professional life have demanded even more effort than was required during expatriation.
> —*American spouse returning from 4-year assignment in France*

Many domestic partners have difficulty finding work after they return home. Sometimes these challenges are related to the loss of professional skills or contacts. Others feel that potential employers may decide not to hire them out of concern that they will leave to relocate in the near future. Still others struggle with establishing the home and helping children adjust, which leaves little time for the job search. Given the potential difficulty of finding appropriate work after repatriation, it is unfortunate that so few multinational firms offer job-finding assistance to partners of returning employees. In the United States, only 2% of repatriated spouses receive such assistance; whereas in Finland, the figure is 15% (Black & Gregersen, 1991a).

Unique Challenges for Japanese Spouses

Japanese spouses (almost all women) face somewhat different circumstances from American or European spouses when they return home. Before, during, and after international assignments, Japanese wives generally have three major roles—that of household manager, mother/educator, and neighborhood member. The role of household manager is an important one that entails handling all of the family's financial matters, including investments. Fifty-four percent of the problems Japanese spouses encountered during repatriation focused on this aspect of their lives. These problems ranged from moving and family finances to housing and living conditions. Generally, these problems stemmed from the difficulty of running a house in Japan after getting used to the better conditions overseas (White, 1988).

In Japanese society, the direction of much of one's adult life is a function of one's education. Because the government and major corporations offer lifetime employment to graduates of Japan's elite universities, much of grammar school, middle school, and high school is geared to doing well on college entrance exams. In fact, nearly all successful applicants to the country's top university have also spent several years attending after-school "cram" courses called *juku*. The mother is the primary contact with schools and teachers as well as the main motivator and coach at home. The popular phrase in Japanese is *kyoiku mamma* or "education mother."

Because the educational system in Japan is so rigid, the problems that schoolchildren have on returning to Japan can exert significant impact on Japanese spouses. These challenges usually result in Japanese women making tremendous efforts on repatriation to hasten their children's assimilation and learning so that the children will not be permanently stigmatized as foreigners and evaluated as educationally impaired.

The role of neighborhood member includes a Japanese spouse's membership in a variety of social groups. Japanese women gain much of their self-identity through these groups and through their ability to belong. Such groups may include mothers who take their preschool children to the neighborhood park each day, flower-arranging clubs, and so on. Overseas, the women usually experience an entirely new set of roles and situations such as accompanying their husbands to company dinners and events, hosting dinner parties in their homes, or attending dinner parties in the homes of their husbands' business associates. In Japan, wives are not involved in these business and social activities, even if they have been involved in them overseas.

The sudden loss of involvement in their husbands' business activities often leads to feelings of frustrations among Japanese women, many of whom have enjoyed their expanded role in their husbands' activities and have gained a new sense of identity and self-esteem.

Impact of Spouse's Adjustment

Most firms do little to support spouses during repatriation. Some executives may feel that firms should not intrude into the family lives of employees, but the reality is that companies expect families to accompany employees on global assignments. Perhaps most important, our research shows that when spouses adjust well to being home, the employee is likely to encounter fewer adjustment problems at work.

STEPS TOWARD SUCCESSFUL REPATRIATION

So far, our discussion has focused primarily on the challenges and dilemmas international assignees and their families face as they try to adjust to life at home. The focus will now shift to what firms can do to help make the process of repatriation less stressful and less difficult.

Define the Strategic Functions of Repatriation

The first step a firm can take toward facilitating effective repatriation is to analyze with the returning employee the strategic functions the employee can accomplish after he or she returns home. Before the assignment, the firm should have clarified the primary purpose for sending the employee abroad: executive development, coordination and control, and/or transfer of information and technology. If the strategic purpose of the assignment was to foster executive development, the assignment after the return should be a critical next step in the development of executive skills and knowledge. If the strategic purpose of the assignment was coordination and control, the assignment after the return could utilize the repatriate's overseas contacts to continue effective coordination and control between headquarters and foreign operations now that more effective relationships have been established. If the strategic purpose of the assignment was to transfer information and technology, the firm should seriously consider what home-country units would benefit from that information and technology transfer. Unfortunately, managers in corporate headquarters often underestimate what home-country units can learn from overseas operations, limiting the probability that effective information and technology transfer will occur.

Unless the repatriate has a clear goal after returning from a global assignment, the investment, usually more than $1 million, to send the employee overseas is likely to be squandered. Furthermore, without a strategic purpose, there are few compelling reasons for the firm to pay significant and systematic attention to the multitude of problems facing the repatriate and his or her family.

Establish a Repatriation Team

Once employees have been told clear and strategic objectives, a team should be formed consisting of a human resource department representative and the expatriate's sponsor. These individuals should initiate preparations for the return at least 6 months before repatriation. If possible, the human resource department representative should have firsthand experience with both expatriation and repatriation. Many of the repatriates we interviewed echoed the feelings expressed in the following comments:

> It would have helped, at least, to have personnel with some understanding of the experience of repatriation. Most of these people have no appreciation of what needs to be done in coming home. Since we have lived internationally and moved back one time before, we knew what to expect and basically had to manage it ourselves.
>
> Have a human resources department that understands the trauma of repatriation. In the best-case scenario, the human resources department would be made up of former expatriates.
>
> —*Finnish repatriate*

> Most people in the personnel department have not had any international experience. Consequently, they cannot understand the process. This is a big mistake.
>
> —*Japanese expatriate*

The supervisor or the sponsor plays an important role on the repatriation team in that he or she is primarily responsible for locating an appropriate position for the repatriate when he or she returns. For example, in the medical system division of GE, sponsors are formally evaluated on their effectiveness in performing this role.

Target High-Risk Repatriates

Once strategic purposes have been defined and the repatriation team is in place, it is critical to offer support to employees who are likely to have more serious adjustment problems. Two characteristics place some employees (and partners) in this high-risk group. First, employees and partners who have been away for an extended period (either on multiple assignments or long individual assignments) are likely to have greater problems adjusting. Second, employees and partners whose international assignments were in a country that is very different from their home country (for example, a German returning from China) will more likely have greater difficulties. These individuals are most likely to have inaccurate perceptions of their home country and company because the two environments are so vastly different.

MANAGE EXPECTATIONS WITH ACCURATE INFORMATION

Because the expatriate, the parent company, and the home country have all probably changed during the global assignment, numerous aspects of the repatriate's perceptions of home may be inaccurate. Firms must therefore manage and mold expectations before individuals come home. This increases the likelihood that expectations will be fulfilled and that the employee will be able to adjust effectively to both work and nonwork issues after repatriation. The importance of managing expectations during repatriation is expressed well in these comments:

> I am Austrian by birth and lived in Germany until I was 18 years old. Then I moved to the U.S., and my ties to Austria and Germany have remained strong all these years. When I had the opportunity to work in England and Germany for the last 2 years, I happily accepted the German part of the assignment. It seemed like going home. When I went back there, I took with me all my expectations about a country that I remembered mostly through holiday visits and through the eyes of my parents and relatives. In contrast, I had no opinion about England before going there—I just went. I went to England first and, much to my surprise, I loved it. I had no real problem adjusting, even though the differences from my life in the U.S. were great. Then came a Germany that I didn't recognize. I found the people very rigid and inflexible. I felt like a foreigner in a country where I had expected to feel very much at home. If I had trouble adjusting, it had nothing to do with the differences between the U.S. and Germany—I expected those. The differences that caused me the most difficulties were the ones between the Germany that I had lived in years ago and the Germany I was returning to.

Provide Sources of Information

Firms have several mechanisms they can use in molding employees' expectations before they come home including sponsors, prereturn training and orientation, home-country visits, and general information.

Designate Sponsors. Sponsors can be a valuable source of information for managers while they are on global assignments and especially just before they come home. Even if an employee has not had a sponsor during the assignment, it is still important to assign one before repatriation. Sponsors generally focus on providing employees with information about company-related changes, but they should be encouraged to also relay more general news about changes that have occurred in the home country.

Several spouses in our study suggested that firms provide "family sponsors" who remain in contact with families both during the assignment and after the family returns home. Family sponsors could be good sources of information about general changes.

Provide Training. Training and orientation sessions are another way that firms can prepare employees and their families for repatriation. As one returning employee said, "Don't just give expatriates a brochure and walk away." Employees need information about changes in their jobs, about how to interact with people, and about changes in the general living environment. Job-related information could focus on structural and political changes in the firm, technological innovations, procedural changes, and so on. Communication-related training could focus on the differences in interaction styles between the country where the employee had the international assignment and the home country. This training could also include warnings that people at home, both at work and in social situations, are likely to be relatively disinterested in the family's experiences abroad. Finally, training and orientation can be provided about housing, changes in compensation packages, tax laws, school systems, price levels, and so on.

Firms should make much more of an effort to provide information to partners and other family members. One spouse from Finland reinforced the importance of providing formal information. After spending 8 years overseas in four different assignments, she stated, "In big companies, it seems so easy to forget that getting through training and orientation is important for the well-being of people when they are overseas and when they come home!"

One question that executives often ask us about repatriation training is "Who will provide the training when employees do not return home in groups?" Clearly, it is expensive and time consuming to send trainers to all parts of the world to provide customized training on repatriation. This strategy might work if groups of employees are coming home from the same part of the world at about the same time. As an alternative, one of our client firms in high technology developed a video-based training program that is sent to employees and their families before they return home. The video provides company-specific information as well as general information relevant to repatriation. In another case, we brought a group of employees from noncompeting firms together for a 1-day program on repatriation. The program was offered at regular intervals throughout the year.

Encourage Home Leave. Along with providing sponsors and training, firms should encourage employees and their families to visit the home country regularly while they are on global assignments and especially just before they come home. Home visits provide international assignees and their partners with opportunities to develop more accurate expectations about what life will be like when they return. As we mentioned earlier, two thirds of the expatriates we studied received paid home leaves from their companies, but only one third of companies required their employees to take home leaves at home. Requiring home leaves to be taken at home can benefit firms in the long run by providing international assignees and their families with the opportunity to acquire important information about home while they are out of the country.

Provide Access to Newspapers/Magazines. Newspapers and magazines from home are another valuable source of information about changes and home-country events. As one spouse from Finland suggested, "When you know you are going back, you should make time to read newspapers and magazines from home. Then you have a much better idea of what is happening." Because newspapers and magazines can be expensive, firms use a number of arrangements. Some send only Sunday editions of home-country newspapers. Others send only weekly or monthly magazines. Still others provide single copies to groups of employees who work together (one problem with this approach is that the newspapers and magazines are unlikely to make it home so that spouses and children can read them). Regardless of the method, the purpose is to provide information about the home country so that international assignees and their families can develop realistic expectations about life after they repatriate.

Preparing the Job Environment

To avoid many of the problems associated with repatriation, the repatriation team, in consultation with the international assignee, needs to explore the employee's career path and options after repatriation. Ideally, the job after repatriation will provide the employee with an opportunity to accomplish one of the three strategic objectives for global assignments; otherwise, firms can expect a low return on the investment in people sent abroad and brought home. The job after the employee returns should also contain some element of challenge as well as a reasonable level of autonomy. Most executives or managers want these attributes in their work, but repatriates are especially attuned to them after working overseas in very responsible positions with high levels of autonomy. In addition to creating challenging jobs with discretion to make things happen, it is important to assess how skills the employee has acquired during the global assignment can be utilized. As one returning expatriate suggested, "Employees who were successful in various worldwide assignments have considerable and varied insight into conducting business. Firms should treat such insight and perspective as an asset rather than discarding, wasting, and hindering such contributions." We concur with this position, as does our research. When the match is appropriate, the employee is much more likely to adjust well to his or her new job. As an excellent example, consider the following executive's experience after returning to Ford:

> After coming home, I took a position that gave me the opportunity to use what I learned while working at Mazda in Japan for the last 3 years. The new job is terrific. Overall, coming home has been easy, since I returned to an area that deals specifically with international activities. In my new group, it is critical to know how Mazda works, and I have that knowledge.

In this case, the job appears to be challenging and utilizes specific skills that the manager acquired overseas. More important to Ford, however, the new assignment enables the employee to accomplish a critical strategic objective—that is, information transfer. This employee was fortunate to have a job with Ford in which he could effectively utilize knowledge he gained during his global assignment, and in turn, the parent company (Ford) was able to benefit from this knowledge.

In some firms and industries, it may not be possible to provide repatriates with ideal jobs because of downsizing, restructuring, and so on. It is better to communicate these situations to employees clearly and early rather than leaving them in the dark overseas. One repatriate from a large U.S. energy firm described what he called a "mushroom-growing" approach:

> Why can't firms provide at least some information about the progress, or lack of it, in finding a new assignment while expatriates are waiting to come home? Correspondence I sent home from Jakarta during the last three months of my stay was never answered. Being in the dark for months is very hard when you know you are going to repatriate!

If placing a repatriate in a good job is not an option, the employee should be told this before he or she returns. Otherwise, the employee will arrive home with inflated, inaccurate expectations that will most likely lead to adjustment problems not only for the employee but for his or her whole family. The "From the Front Lines" in this chapter illustrates why it is important to keep the lines of communication open between employers and repatriates and not lose sight of the value of international expertise to the firm.

In addition to choosing an appropriate job for the repatriate, the firm should consider the backgrounds of the repatriate's new supervisor and coworkers. These people often have little or no international experience and have difficulty understanding and working with someone who is having a difficult time adjusting. Among American repatriates, very few (29%) have supervisors with any international experience (Black, 1991). Generally, supervisors with no international ex-

FROM THE FRONT LINES:

"Motorola: Expectations Versus Reality After Repatriation"
by John C. Murphy, Director,
Global Transpatriate Rewards, Motorola Inc.

When an employee returns from an international assignment, both the employee and management often hope and plan to find a position for the employee in which his or her newly acquired international experience will be fully utilized. All too often, though, reality gets in the way in the form of economic downturns, business downsizing, reorganizations, and the associated elimination of jobs. One employee's experience at Motorola serves as an excellent example of the reality of the job situation after an employee returns from a long-term assignment abroad.

When David (not his real name) was repatriated, he had worked for a total of 9 years in Geneva, Switzerland, and Manila, Philippines, as a regional HR director. On his return to the United States, he was placed in charge of Motorola's U.S. compensation program requiring no global or regional involvement or accountability. David performed in this role for approximately 5 years, all the while wondering if his knowledge of international issues would ever be utilized. Then his direct supervisor retired, and David was promoted to a position requiring responsibility for the total global compensation/benefits program for a major segment of Motorola's business.

After 6 years in that job, David's position was split into two functions. Although he was given his new corporate position during an economic downturn, David and his new position made for an ideal match. In charge of Motorola's U.S. and international relocation policies, vendor management, competitive assessment, and cost management, David could finally take full advantage of his vast knowledge of issues affecting global operations. Equally important, Motorola could benefit from his extensive international experience.

What is to be learned? Although David's initial assignment on repatriation was not one he would have chosen, and Motorola did not profit immediately from David's experience working abroad, the organization never lost sight of his skills and abilities, and over time, he was placed in positions in which both he and Motorola benefited.

International managers cannot always pick and choose the timing of ideal matchups. However, with some serious organizational planning, both the employee and the organization can benefit in the end.

perience can be expected to have little understanding of or empathy for the challenges the repatriate is experiencing (Paik et al., 2002). In some cases, supervisors and coworkers have actually inhibited the adjustment process. An American expatriate confided, "When I came home, coworkers were very jealous of my [new] assignment, even though my responsibilities were vastly decreased in comparison to my recent overseas position." In Finland or Sweden, coworkers and supervisors may be less jealous of the employee's new assignment but more jealous of his or her new tax-free automobile and tax-free earnings because cars and earnings are heavily taxed in Scandinavia.

Taking a different perspective, one American expatriate returning to a firm in the U.S. transportation industry said, "I was shocked at the animosity of coworkers because I had learned to work successfully with the Japanese during my international assignment." After experiences like this, repatriates learn to keep their mouths shut about their international expertise.

Coworkers and supervisors at home can frustrate the strategic purposes of a global assignment and return position, as happened in this American expatriate's case: "There was a lack of knowledge of and interest in what is happening in the world outside this company. No one really cared what I had done or learned overseas. It seemed like everything had to be 'homegrown' in the U.S.A. or in the parent company." To avoid these potentially significant problems, we suggest that firms provide training and orientation not only to returning employees but also to coworkers and supervisors.

Providing Appropriate Compensation

Most expatriates experience a significant decrease in their overall compensation after returning from a global assignment. This decrease often reflects the rewarding A while hoping for B approach to compensation discussed in chapter 8 (this volume). This focus on offering monetary rewards as incentives to accept international assignments often sets firms and employees up for failure when the employees return because most international assignees suffer significant financial losses on returning. Although it is easy to justify providing increased financial support for an employee in a faraway country, compensation specialists can more easily gauge financial need when the employee returns. Perhaps most important, firms often do not recognize the need for extra compensation or assistance during the repatriation period because coming home is certainly not seen as a hardship worthy of perks.

Nevertheless, firms should still pay attention to the compensation package it offers repatriating managers; otherwise, these employees will have more difficulty adjusting to work and home after repatriation or will leave the firm and go to work where they feel that their international expertise is valued. A first step in the development of an appropriate repatriation compensation package is to compare the repatriate's compensation to that of other employees at home. Unless firms make such comparisons, problems can occur, as this American expatriate found out:

> During the international assignment, I went through hell getting any recognition for my performance from my direct supervisor stationed back home. After returning to the U.S., my salary, which was not adjusted for the 2 years I was gone, was terrible. When I got back, I had to work my tail off to regain parity. Now the job is done, and life goes on.

To prevent compensation-related problems, firms should follow the three-step approach to compensation outlined in chapter 8 (this volume).

Locating Adequate Housing

Finding adequate housing can be a major challenge for repatriates, although some repatriates have positive experiences:

> Coming back to our own home really helped. We had a place to identify with, and friends and neighbors who cared about us. It also helped that our kids were in the same school when they returned. We were in contact with teachers during the assignment, and they remembered us.
> —*American spouse*

> It is critical to keep your previous flat in your home country, especially when teenage friends and social contacts are so important. It was terrific to have the children return to their own school and neighborhood.
> —*Finnish spouse*

The best-case scenario is for the repatriating employee, partner, and children to return to the same home or at least to the same neighborhood as they lived in before they went abroad. As mentioned previously, employees who rent their homes may have to repair damage caused by tenants. Firms need to consider whether they should provide temporary housing while house repairs are being made, which, depending on the extent of the damage, may entail major expense for the family.

People who sold their homes or did not renew leases for apartments before leaving on global assignments will need assistance in locating, purchasing, and moving into new accommodations. House-hunting trips during the last few months of a global assignment can be very helpful. If the house-selection process must occur after the employee returns home, the firm needs to provide adequate time for the employee and his or her partner to find appropriate housing.

Providing Support Groups

In our research, repatriates and partners often suggested that firms should consider providing informal opportunities to meet and socialize with other repatriates and their families. As one spouse said, "Support groups could help us answer the many questions a returning family has. A great many changes occur during the global assignment, and searching for the answers alone can be most frustrating and cause needless tension in the family unit." This is an inexpensive endeavor for employers and has the potential to provide important benefits.

Planning for "Downtime"

The challenges of coming home to new work and home routines often require significant amounts of time, but many repatriates fly home one day and go back to work the next. Consider the following comments made by Americans:

> Arrived home on Saturday. Started work at 150% on Monday. I have worked constant 70-hour weeks since.

> I arrived home on Tuesday and started work on Wednesday and have been working 10- to 12-hour days since. I haven't adjusted yet to much of anything and really feel depressed.

> I worked 14-hour days, 6 days a week. There was little time to look for housing, yet the company still pressured me to move out of a hotel, in order to get me off the "expense" status.

Many European repatriates suggested taking a 3- to 4-week vacation after returning from a global assignment to get things in order. The Americans recommended taking 2 weeks at most. Some repatriates suggested that firms should force returning employees to take time off. Regardless of the amount of time the employee takes, the important point is that firms should allow repatriates time to adjust to being back in their home country.

Appreciating Contributions to the Firm

Finally, firms need to let their global employees know how much they are appreciated. Several spouses told us how much it would have meant to them if their spouses' firms had expressed their appreciation for the job their spouses did or demonstrated a little more interest when they came home from abroad. One spouse from Finland said, "Why can't companies take a moment to say 'thank you' to the wife and children for the sacrifices they made to uproot, go overseas, and come back home?" We agree.

END WITH THE BEGINNING IN MIND

As the last step in a global assignment, effective repatriation provides positive feedback to the next generation of global employees. When repatriates' coworkers are offered global assignments, they may say what this Ford employee said: "Were we supposed to have difficulties? Things went very well!" How had Ford treated this employee and his family after his 3-year assignment in Japan? Ford provided 3 hr of general and culture-related training, which was seen as valuable. The company also placed the employee in a position with clear work expectations and moderate levels of responsibility as well as an excellent financial package (at least from the repatriate's perspective). The individual returned to the unit he worked in before he left for his global assignment with coworkers and supervisors who placed a high value on his global experience. Clearly, he had accurate expectations about coming home.

The bottom line is that inattention to the difficulties of repatriation has a negative impact on employees' performance and thus on company performance. By contrast, small and often relatively inexpensive steps can lead to significant returns and enhanced competitive position in a global marketplace.

CHAPTER
10

Retaining: Utilizing the Experienced Global Manager

In chapter 9 (this volume), we discussed the process of repatriation and ways firms can increase the chances that returning employees and their partners will make a smooth adjustment to being home. In this chapter, we shift our focus to the dynamics of organizational commitment and various factors that can affect such commitment.

A successful adjustment to living and working in the home country is critical to high-level job performance after global assignments. Likewise, organizational commitment is critical to keeping high-performing repatriates in the firm (Lazarova & Caligiuri, 2001).

GLOBAL MANAGERS AS STRATEGIC ASSETS

If a multinational firm invested between $2 million and $4 million in a piece of critical production equipment over the past 3 years, it would be hard to imagine the production manager not taking serious action if the equipment were headed out the door to a competitor's production facility; yet each year executives and managers (in whom firms have invested millions of dollars) walk out the company door after returning home from global assignments. Although less easy to quantify, the costs to the company of losing these human "assets" can be as significant as letting that piece of production equipment go out the door.

According to one study (Black & Gregersen, 1999) of more than 750 companies, 25% of managers leave their companies within a year after completing a foreign assignment. This is twice the rate for managers who have not worked abroad. This problem is not unique to U.S. companies either. As just one example, in a study of German repatriates, 51% said they were willing to leave their company for another organization, whereas only 25% said they were unwilling (Stahl, Miller, & Tung, 2002).

Even if repatriates stay in their parent firms after coming home, most feel their market knowledge, technical skills, foreign-language ability, and so on are underutilized. These sobering realities have compelled leading multinational firms to look for answers to how they can better utilize their global managers' skills and decrease the turnover rate among managers returning from global assignments. In the rest of this chapter, we discuss some of the issues companies must address if they hope to retain and take advantage of the considerable expertise of their global managers.

RETENTION PATTERNS IN MULTINATIONALS

We have found that four general patterns of behavior occur during the repatriation process (Black & Gregersen, 1992). These patterns are shown in Fig. 10.1. Ideally, multinational firms want to achieve functional retention—that is, for their repatriates both to be high performers and to stay in the firm.

Two factors are closely associated with whether a firm achieves functional retention: (a) its repatriates' levels of adjustment and (b) its repatriates' commitment or loyalty to the parent company. As discussed in chapter 9 (this volume), high levels of adjustment lead to high job performance. Furthermore, high commitment to the organization after repatriation leads to strong intentions to stay with the firm.

		Organizational Commitment	
		High	Low
Repatriation Adjustment	High	Functional Retention High performance High intent to stay	Dysfunctional Turnover High performance Low intent to stay
	Low	Dysfunctional Retention Low performance High intent to stay	Functional Turnover Low performance Low intent to stay

FIG. 10.1. Repatriation outcomes.

Functional Retention

Firms want to achieve functional retention for very strategic reasons. Most obviously, as we have pointed out in several different places throughout this book, international assignments are the most powerful means of developing future global leaders. Managers who performed well overseas and are doing well on their return have the highest potential of being global leaders in the future. Unfortunately, in many firms, the percentage of managers who fit this description is much too small.

Dysfunctional Turnover

More unfortunately still, in many firms, the percentage of managers who end up in the dysfunctional turnover category is far too large. Repatriates who fall into this category have adjusted well to work, to interacting with people in the home-country, and to the general home-country environment but do not have a strong commitment to the parent company. The loss of these high performers usually represents a significant loss of resources including firm-specific experience, global perspectives, international skills, and so on.

Firms lose not only their investment when high-performing repatriates quit but also the chance to accomplish strategic objectives. For example, consider the consequences to an American firm when it sends an employee to Japan to gain leadership experience and insight into the Japanese market. If the employee quits after coming home, the losses to the company may be significant. It will be unable to benefit from either the manager's leadership abilities and experience or from valuable insight the manager may have gained about the Japanese market.

There are many reasons high-performing repatriates quit. Many leave because they feel they are unable to utilize their global experience in the position they are offered after repatriating. Another explanation gaining prominence is that some employees who accept global assignments place a high intrinsic value on the experience rather than viewing it as furthering their careers with their current employers (Stahl et al., 2002). These employees are likely to leave if the job they hold after repatriating does not capitalize in some way on the growth that took place while they were abroad.

Functional Turnover

Because it is impossible to be 100% correct in selecting managers for international assignments, some functional turnover is to be expected. In other words, most firms are likely to have an occasional employee who neither adjusts well nor performs well after coming home. However, if a large proportion of a company's repatriates fall into this category, serious corrective action is needed, probably in the selection process. High functional turnover is a clear sign that the wrong people are being sent on global assignments.

Dysfunctional Retention

Dysfunctional retention—retention of low performers—is obviously not desirable for a firm. Repatriates who fall in to this category generally have high loyalty to the company but fail to adjust effectively to their new work, social, and general environments after returning from abroad. This pattern can occur when (a) the wrong people are sent on global assignments, (b) they are not properly trained and perform poorly, and (c) they have no other job prospects after repatriation. If most of a firm's returning managers fall into this category, serious intervention is needed before international operations get a reputation in which the "has beens" or "never will bes" get dumped.

FACTORS AFFECTING ORGANIZATIONAL COMMITMENT

For firms to achieve the desired goal—namely, retaining high-performing employees (functional retention)—they need to first focus on accomplishing two critical objectives: ensuring that employees returning from global assignments adjust well during repatriation and maintain or increase their commitment to the organization.

The first step toward sustaining a sufficient level of commitment after repatriation is to monitor commitment patterns during global assignments before employees return home. In chapter 6 (this volume), we discussed how employees can become committed not only to a parent company (which sent them out on a global assignment) but also to a foreign operation during the assignment. In addition, we presented some of the specific factors that determine whether a firm develops and sustains commitment while employees are abroad. For example, research has shown that employees with several years of experience in the parent firm and who have clearly defined but fairly autonomous jobs overseas tend to be more committed during global assignments. When employees return, firms need to either redevelop or sustain these employees' overall commitment to the organization.

The specific factors that influence this critical commitment to the parent firm are somewhat different from those that play an import role in loyalty during the assignment. We have identified four general categories of factors that increase or decrease employees' level of commitment during repatriation: individual, job, organizational, and nonwork factors. These factors are displayed in Fig. 10.2.

Individual Factors

The company may not be able to control individual factors, but being aware of their impact allows for better planning and anticipation regarding such matters as which managers might have more difficulty or ease in adjusting during repatria-

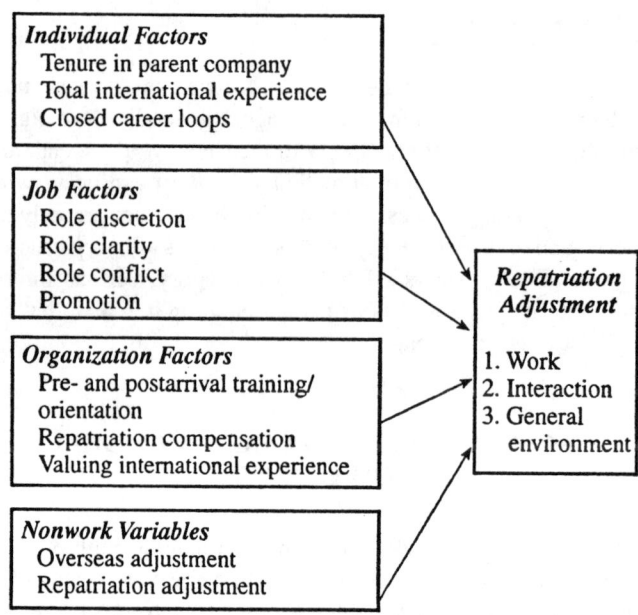

FIG. 10.2. Basic framework of organizational commitment during repatriation.

tion and which ones might have more or less loyalty to the company and stay or quit after returning home.

Tenure in the Parent Company. Most employees have extensive experience in the parent company when they embark on a global assignment. For example, of the hundreds of managers we studied, most had at least 12 years of experience in the parent company. Having made this investment of time and energy tends to bind the employee to the company or increase the sense of loyalty and commitment. This is true whether the employee is working at home or abroad.

Tenure in the parent company is also an important factor in sustaining commitment to the parent company during repatriation. However, in the case of Japanese and Finnish repatriates, tenure in the parent company appears to be unrelated to loyalty. The insignificant impact of tenure on commitment for Japanese or Finnish expatriates may be partly a result of the comparatively low job mobility across companies in both Japan and Finland.

Total International Experience. Another important individual factor relevant to commitment during repatriation is the total amount of time individuals have worked outside their home country. How much international experience do employees on global assignments have? This question is important because those employees with the highest levels of international experience (i.e., career expatriates) have usually made significant investments in their international careers. For this reason, some managers stay out of their home country and continue to utilize their international skills and knowledge in other assignments abroad. Furthermore, some employees develop a relatively high level of disinvestment in their home country while developing their valuable international skills. Expatriates with extensive international experience can develop an aversion, almost an "allergic" reaction, to their home country and to socializing with home-country nationals. As one American career expatriate in the banking industry explained, "There is little value placed on international experience within my company. Home nationals tend to think that the U.S. is so different that anything learned overseas is largely irrelevant to the domestic U.S. market."

Japanese career expatriates expressed similar sentiments:

> When you come home, no one appreciates your international experience as much as you do, or as much as the people you worked with overseas do. In fact, overseas, the mere reality of it, does not even exist for most "domestic" employees in this company. My suggestion for international firms? They could at least brief or work with the people at home who have never been overseas and help them appreciate an international perspective a bit more.

Reflecting the problems expressed in this man's comments, our research has found that American expatriates with high levels of international work experience are generally less committed to their parent companies after returning home. In contrast, Finnish expatriates with high levels of international experience tend to be more committed to their parent firms. This relationship may result from the newness of international activity for Finnish firms as compared to American multinationals and from the importance Finnish firms place on international experience.

Movement Pattern. A final individual factor that can play an important role in sustaining commitment to the parent company during repatriation is the pattern of assignment before, during, and after the international posting. Some expatriates refer to these patterns as *career loops.* In a closed career loop, individuals work in the same division of a firm before, during, and after a global assignment. In an open career loop, divisional continuity is broken as employees are shifted from one division to the next. In our research, between 30% and 40% of expatriates from the United States, Japan, and Europe experienced closed career loops throughout their international assignments.

Creating closed career loops is a way to enhance company commitment in that it ensures that employees returning from global assignments will be working in units and with people who are already familiar to them. The importance of career loops was reflected in our research, which found them to affect expatriates' commitment during repatriation. Interestingly, Japanese expatriates are an exception. Closed career loops do not seem to affect the commitment of Japanese during repatriation. This may be a result of the routine transfers these managers make across functional and divisional boundaries, a practice that is not as common in the United States or Western Europe.

Job Factors

As we discussed in the context of adjustment, the nature of the job a manager has after returning home is one of the most powerful factors affecting commitment to the organization and resulting intentions to stay or move on to a different firm.

Expatriates usually become accustomed to holding significant, visible jobs during global assignments. Typically, these jobs have high levels of autonomy and discretion. Moreover, managers on global assignments often develop a strong action orientation and a feeling of responsibility for making things happen. After returning home, however, these managers often end up in temporary low-level holding-pattern positions that fail to utilize their skills. Unfortunately, their overall job dissatisfaction leads them to start looking for "real work." These comments are representative:

> It is very important to know what the assignment will be after returning home. Companies will easily lose their best expatriates if the work is not challenging and interesting enough.
>
> *—American expatriate*

> My current position has very little to do with what I learned overseas. What a waste for this company, to spend all the money to send me overseas and then bring me back and not even utilize what I learned while I was there.
>
> *—Japanese expatriate*

> My suggestions for other expatriates coming home? Update their resumes. The experience that you gain overseas is seldom valued and often discounted when you come home. Other companies will pay for what you know and can do for them. I can't understand why expatriates and companies cannot learn to plan ahead and make use of the skills and experience gained overseas. The rule seems to be that return assignments are just "landing places." What's lost is the opportunity to use an expatriate's ability to the fullest benefit of both the company and the employee.
>
> *—Finnish expatriate*

Supporting these comments, our overall research found that when American, Japanese, and Finnish firms provided repatriates with clearly planned jobs, significant levels of discretion, and few conflicting messages about how to do their work, repatriates exhibited higher loyalty to the parent company.

Organizational Factors

Several human resources policies and practices can enhance organizational commitment during repatriation. Each of these activities communicates to returning employees that the firm cares, is supportive, and—perhaps most important—can be counted on when it sends and retrieves employees to and from international postings.

Repatriation Training. Firms can take an initial step toward being dependable and supportive in the repatriation process by providing training and orientation to employees and their families both before and after they return home. This training not only facilitates cross-cultural adjustment during repatriation but also communicates to returning employees that they have not been completely out of sight and out of mind while they were away. In chapter 6 (this volume), we discussed the positive effects that training can have on commitment to the parent company during a global assignment. Unfortunately, we cannot report strong data on repatriation training because so few firms provide it. In fact, 97% of American, 94% of Finnish, and 98% of Japanese expatriates receive fewer than 4 hr of training and orientation on or before coming home.

Compensation. In addition to prereturn and postreturn training, providing adequate repatriation compensation is another way firms can communicate their support. Many of the challenges of repatriation have already been outlined in chapters 8 and 9 (this volume). The more firms can ease the financial burdens of settling into life in the home country, the more likely repatriates are to exhibit commitment to the parent company. From our research, we know that this correlation is especially true for Americans returning home. The most troubling aspect of repatriation compensation is the potential loss of income and the associated drop in living standards. To minimize the financial loss of coming home and to maximize a repatriate's organizational commitment, firms should follow the three-step approach to financial compensation (outlined in chap. 8, this volume), which minimizes the income shifts throughout the assignment cycle and reduces feelings of financial inequity after the return home.

Value of International Experience. The final and most critical factor contributing to commitment during repatriation is the extent to which a firm communicates to employees throughout the company that it values international experience and perspectives. This message is communicated through the job the employee is given after returning home, the attitudes of immediate supervisors and coworkers, the formal and informal reward systems for employees with international experience, the promotion paths, and so on. All of these factors collectively create a perception that firms either do or do not value global experience.

The most important factor in sustaining repatriates' commitment to the parent company is developing an organizational culture that values international experience and expertise. Unless firms can foster such a culture, they are likely to create repatriates who feel unappreciated and go elsewhere for the recognition they feel they deserve. Unfortunately, the vast majority of expatriates (between 60% and 80%) feel that their firms do not genuinely value international experience. This low utilization rate is reflected in these expatriates' comments:

> Your own perceived increase in "value" after an international experience is just that—perceived. Expatriates should be prepared for the fact that few—precious few—people are interested in their international experience.
> —*American expatriate*

> The most challenging part of coming home was the fact that my overseas experience was not valued in my company as I had expected it to be. Keep your expectations very low when coming home. Remember that no one really cares that you spent time overseas.
> —*Finnish expatriate*

> Look at your international experience as an extremely valuable asset, but recognize that few others will fully appreciate it.
> —*Japanese expatriate*

Leading firms retain committed and loyal expatriates by carefully assessing the culture in the company and reinforcing the value of global experience and perspectives throughout the organization. Although this prescription may seem simple, many firms are trapped in the dilemma of rewarding A while hoping for B, at least from the repatriates' perspective. For example, American executives are well known for touting the value of international experience. Yet, in most companies, previous international experience is still not a critical selection criterion in executive promotion decisions. These and other incongruities between what firms say about the importance of international experience and how they actually treat employees who have it must be examined if firms hope to effectively and consistently communicate that global managers with global experience are valued.

Nonwork Factors

The final category of factors relevant to sustaining commitment during repatriation has to do with cross-cultural adjustment during and after the international assignment. Past research (Stroh, Gregersen, et al., 2000) has found that the more adjusted American expatriates were to the foreign culture, the less committed they were to the parent company; however, this adjustment had no impact (either positive or negative) on commitment during repatriation. In contrast to Americans,

Japanese and European expatriates who had adjusted well during the international assignment were actually more committed to the parent company after the international assignment.

This positive impact may be a function of at least three factors. First, research has found that Japanese and European firms pay more systematic attention to expatriation policies and practices than American firms do. This planning may result in more parent-company support during the global assignment, which translates into more effective adjustment and more significant commitment. Second, Japanese and European repatriates are simply less mobile than their American counterparts. As a consequence, Japanese and Finnish expatriates who adjust well overseas are likely to offer a positive return on investment of time and energy by staying with the parent company rather than attempting to change firms in relatively immobile labor markets. Third, Japanese and European firms generally place a higher overall value on global experience than do American companies. The global experience counts more, so it pays for employees to adjust effectively during global assignments.

After returning home, expatriates face many significant challenges. Many of these challenges were discussed in chapter 9 (this volume). Here, we reinforce the importance of effective adjustment after repatriation. In particular, effective adjustment to the home country after global assignments positively influences commitment to the parent company. Essentially, efforts that individuals and firms make to facilitate the repatriation process not only help repatriates adjust and perform well but also encourage commitment to the parent company because repatriates feel the company can be counted on for support throughout the entire global assignment cycle.

STRATEGIES FOR SUSTAINING COMMITMENT

Three key issues influence commitment during the process of repatriation. First, firms can increase commitment by creating a culture that genuinely values international experience, perspectives, and skills. Second, firms can retain high-performing repatriates by carefully planning appropriate return assignments. Third, firms can reduce dysfunctional turnover by paying particular attention to high-risk repatriates.

Creating an Organization That Values Global Experience

Creating an organizational culture that communicates the positive value of global experience is the most critical factor that we have identified for developing commitment to the parent company after returning from a stint abroad. A global assignment demands personal and familial sacrifices. Firms should ensure that those sacrifices are honored. The last thing that international firms need is repatriates who suggest, as one did to us, that future expatriates "must realize that whatever accomplishments were made overseas, they do not count for anything back at headquarters. You have to earn your 'stripes' all over again, so look after yourself, because nobody else will back home." Firms need to pay attention to the many formal and informal activities that collectively create the perception that firms value or do not value experience gained abroad.

Employees notice what is and is not rewarded in their firm. In particular, promotions and executive advancement are critical conduits of information about what really counts. Many firms throughout the world tout international experience as essential to global expansion, yet their own boardrooms and executive ranks contain no one with experience working outside the United States. By including international experience as a central criterion for advancement and promotion, firms can take an essential first step in communicating the value of international experience. The difference in attitude is apparent in companies such as 3M, Citicorp, and Colgate in which more than 75% of senior managers have international experience. In these and other leading-edge companies, people know that global business savvy is valued.

Compensation policies are another significant way that employees learn whether a firm "puts its money where its mouth is." If employees returning from global assignments do not experience significant financial losses, other employees are likely to sense that international experience is valued. If the compensation policies outlined in chapter 8 (this volume) are used throughout the global assignment cycle, repatriates are less likely to experience feelings of inequity. Basically, ensuring that returning employees are compensated equitably is an essential link in creating an international culture.

Company-provided training before and after repatriation not only facilitates adjustment to reverse culture shock but also shows repatriates that the firm is aware of and pays attention to the challenges of coming home. When a firm provides no training or orientation during repatriation, the implicit message is that the firm is complacent ("We don't care about your problems") or ignorant ("We aren't aware of any problems"). In both cases, repatriates quickly learn that their international experience has little value to the firm.

An appropriate return job assignment also communicates that the firm genuinely values international experience and perspectives. If repatriates are assigned to positions that fail to utilize their international skills and knowledge, they quickly learn that these skills and knowledge are unimportant. As one American repatriate put it, "If a firm really wants to go global, it should value international experience by actually using expatriates' skills and experience gained overseas and by putting more effort into planning the reentry position and career progression within the firm."

In addition to formal mechanisms, firms can communicate that they genuinely value global experience in a number of informal ways. These informal mechanisms cannot be easily "legislated" (by instituting repatriation training or compensation). Instead, these mechanisms often result when internal human resources policy statements are put into practice. For example, Citicorp genuinely values international experience and promotes individuals who have completed global assignments. As a consequence, board meetings, executive discussions, and strategic planning processes are permeated with international rather than just home-country perspectives. Seeing global executives regularly raise international issues communicates to repatriates that global experience really does count.

Another way repatriates pick up on whether a company values global experience is by observing how many "foreigners" are working in the home-country office. As company headquarters becomes globalized, repatriates are more likely to look around when they return and notice the global environment.

Finally, the war stories traded among employees about the benefits and costs of global assignments are powerful indicators of how much individuals think firms value global experience. Are the informal comments about so-and-so losing his job, marriage, and career after going overseas or about employees who accomplished strategic objectives and returned home to a company that valued their accomplishments? These examples are a few of the numerous informal ways in which repatriates are tipped off about the real value of global experience within a firm.

Strategically Planning the Return Position

Commitment to the parent firm is partly a function of having a job in which one feels a strong sense of responsibility. If the positions to which repatriates return are unclear or trivial compared to what they experienced overseas, repatriates are less likely to feel a sense of commitment to the jobs or to the companies they work for. Consequently, firms should identify strategic purposes that expatriates can accomplish after coming home, such as continued executive development, coordination and control, or transfer of information and technology. The strategic purpose and how the job facilitates its accomplishment must be clearly and personally communicated to the individual.

A repatriation team, consisting of an organizational sponsor and a human resource representative, should examine potential jobs within the firm beginning several months before the employee comes home. The firm should consider the level of autonomy and discretion that the person has had during the international assignment. If possible, the return position should have an equal or greater level of challenge and job autonomy; well-planned and well-selected return positions can be critical to a returned manager's sense of responsibility in the job and sense of commitment and allegiance to the firm. If a responsible position cannot be created at home, the firm should communicate in advance—before repatriation—that the return assignment may be less than desirable; accurate expectations are especially critical when individual expectations may not be met. Finally, firms should attempt to utilize specific international skills that the repatriate has acquired during the assignment, including language, negotiation, and cross-cultural communication skills or essential knowledge of the foreign market.

Although most firms assume that a good return job will turn up—almost magically—leading firms assume exactly the opposite. They start with the assumption that unless both the individual and organization expend extraordinary effort, a good match is not likely. They know that good matches are made; they don't just happen. As a result, these firms invest significant time and energy into matching the manager to the return job. As the "From the Front Lines" for this chapter illustrates, both employees and organizations clearly benefit when they work together to find the best jobs for repatriates.

Roughly 3 to 6 months prior to repatriation, Monsanto employees conduct a personal inventory of their international assignment experience. This inventory includes four components:

1. A review of the developmental objectives of the assignment that they set prior to going overseas.
2. A written list of how their career interests have changed and what direction or possible directions they see themselves going.
3. A summary of the new knowledge and skills the employee acquired during the assignment.
4. A summary of all this information and a description of the type of job and unit in the organization where the employee feels he or she could make a contribution and continue to develop.

Concurrently, a team composed of the employee's line-management sponsor and a human resource department representative should conduct an assessment. One key to this team's effectiveness at Monsanto is that team members generally have completed international assignments themselves. In most companies, only a small percentage of employees have such experience; for example, only about 10% of line or HR managers have been on international assignments. At Monsanto, the repatriation team conducts a broad review of impending or possible moves and job changes over several months.

The second step requires a careful look at the individual's personal assessment and a discussion with that individual about what he or she sees in the future. The third step is the "matchmaking" process: determining where the individual's capabilities and desires best fit organizational opportunities and needs.

Obviously, timing is not always perfect, but Monsanto has drastically reduced its repatriation turnover due to the deliberateness of its process. The Monsanto process produces better matches and action all the way around, and it engenders feelings in the repatriating manager of being valued and treated fairly even when ideal matches are not possible.

FROM THE FRONT LINES:

"Leveraging International Experience for Optimal Advantage"
by Amy Glynn, Global Human Resources Consultant,
Dow Jones & Company

Given the huge investment most companies make in employees they send abroad, retaining returning employees and leveraging their international knowledge after they are repatriated makes tremendous sense. However, as Ken Herts's story reminds the reader, by maintaining contact with the home office and being flexible, employees can greatly increase the chances that the value of global experience will be recognized.

Ken Herts was a senior manager with Dow Jones's market-data division when he found out that the division was being sold. Basically, Ken had two options: He could join the acquiring company, or he could find a new slot within Dow Jones.

Ken discussed his goals with senior executives in New York, and soon he was offered the position of publisher at *Wall Street Journal Europe*. In his new job, which would take him to Brussels, he would be responsible for the paper's business activities.

During his time in Belgium, Ken maintained his contacts with headquarters. He also kept communication open with executives in New York.

After several years, Ken decided it was time to repatriate. Together, Ken and his manager worked to identify opportunities in New York. After the company found a local hire to replace Ken, he found a staff position in New York. It was not a step forward in his career, but Ken took it to facilitate a move.

Within a few months after his return, Ken learned that the Dow Jones Newswires group was looking to hire a vice president and general manager. Ken's international experience in the print division made him a desirable candidate and he was offered the job.

As Ken puts it, "I still benefit from knowledge of the players in overseas markets and contacts I have in those markets. Newswires is a global business, and when dealing with companies in other countries, I personally know many of the players, and my time spent in Europe gives me more credibility with them."

Ken's successful transition allowed the company to take advantage of the investment it made in Ken's overseas experience. Equally important, because Ken maintained contact and was flexible during and after his global assignment, he was able to land on his feet.

Identifying and Tracking High-Risk Repatriates

Firms can better manage their repatriation results by profiling and tracking high-risk repatriates. Specifically, repatriates with low tenure in the parent company or many years of international work experience are least likely to show commitment to the parent company after returning home. Interestingly, the same is true for employees who go abroad. Accordingly, both during and after global assignments, firms should pay special attention to repatriates who have been with the firm for only a few years. However, this must not occur at the risk of neglecting employees with long tenure who are most likely to detect incongruities between corporate talk and corporate action regarding the importance and relative value of international experience. This point is critical because these repatriates often have international experience that could strategically benefit either corporate headquarters or a domestic operation with international linkages. Firms also need to pay special attention to selecting appropriate return assignments for high-risk expatriates whose identity and sense of self-worth are connected to their global skills and experience. If these skills are not utilized in their return assignments, these valuable human assets may leave.

CREATING COMPETITIVE ADVANTAGE BY KEEPING THE BEST GLOBAL MANAGERS

In the best-case scenario, multinational firms send people on global assignments to accomplish strategic objectives such as executive development, coordination and control, or the transfer of information and technology. These objectives can be furthered if repatriates are able to accomplish similar objectives in their return assignments. However, these objectives cannot be accomplished if repatriates do not perform well and leave the firm soon after returning home. More generally, multinational firms that make a commitment to effectively utilizing their repatriates' global knowledge are much more likely to create sustainable competitive advantages than companies in which global managers are utilized less effectively. All types of firms can gain a competitive advantage by keeping the best global managers after repatriation, but the advantage is especially pronounced in companies at higher levels of globalization.

At the export stage of globalization, firms send fewer employees overseas and thus have fewer repatriates. Nevertheless, the large investment made in each global manager and the importance of each repatriate to the future international expansion of the firm suggests that retaining the best global managers is critical even for export firms with fewer employees who are working abroad.

Firms at the multidomestic stage of globalization send more employees abroad than export firms do. However, in contrast to multinational and global firms, multidomestic firms tend to keep repatriates in individual assignments or send them on multiple assignments without bringing them home. Thus, compared with multinational and global firms, multidomestic firms have fewer needs for their repatriates. When multidomestic firms do repatriate international managers, they must do it effectively to accomplish current strategic functions and, more important, to facilitate future international expansion because a multidomestic firm may develop into a multinational or global enterprise.

Multinational and global firms have the largest numbers of employees moving to and from the host and home countries. The issues addressed in this chapter and in chapter 9 (this volume) are very relevant to multinational and global firms because these firms have made the most significant investments in international managers and executive assets. One financial institution we work with whose total annual profit is approximately $150 million invests close to $30 million per year in salaries and support for its overseas managers. These significant costs reinforce the need for multinational and global firms to strategically and systematically institute repatriation policies and practices aimed at encouraging their best managers to stay with the firm.

Multinational and global firms need people in global assignments to serve vital strategic functions, but they also need these people to bring back international perspectives, market knowledge, technology improvements, and so on. If the repatriation process is mismanaged, repatriates will give counterproductive advice to those still on assignment:

Advice about coming home? Don't.
—*Japanese expatriate*

If you like your overseas position and do not have the desire to work in a bureaucratic corporate environment, stay where you are and enjoy it. If I had the choice all over again, I would go right back to Germany and live there as long as possible.
—*American expatriate*

Unfortunately, the result of this advice could be that fewer and fewer employees would want to come home, whereas multinational firms need them to return. Unless they return, "global sclerosis" can set in. Several repatriates described this phenomenon, which occurs when a company has only a few repatriates:

I was shocked at the severe lack of worldwide thinking in the executive suite.
—*American expatriate*

My coworkers ridiculed me for carrying and reading *Newsweek* at the office after I had spent 4 years in the United States. They made me feel like a traitor to my company.
—*Japanese expatriate*

> When I returned home, I found that the home organization was totally unaware of how business is really transacted overseas.
> —*Finnish expatriate*

Firms at any stage of globalization need to send people overseas and manage their repatriation effectively. Otherwise, strategic objectives will not be accomplished. The importance of managing international assignments becomes even more crucial when firms are at higher stages of globalization and more global managers are necessary to accomplish strategic objectives. Moreover, firms can make more effective transitions between stages of globalization, from an export stage to a more integrated global stage, for example, if they have managed previous global assignments well. Under these conditions, key decision makers will have the necessary global perspective and experience and other managers and employees will want to accept assignments overseas. Thus, keeping the best international managers beyond repatriation is valuable not only because it helps companies accomplish global objectives today but because it ensures that the company will have high-performing employees who will want to pursue global objectives in the future.

CHAPTER
11

Managing the Entire Global Assignment Cycle: Establishing Best Practices

Most people are so hardheaded that they need a real smack on the head before they are willing or able to rearrange their mental maps of the world. Executives do not generally receive in-depth international management training in master of business administration (MBA) programs, in in-house executive education programs, or from their work experience (Porter & McKibbin, 1988). Less than a quarter of the CEOs in America have gone on an international assignment (Carpenter, Sanders, & Gregersen, 2001; Gregersen et al., 1995). Of the CEOs who have worked abroad, their most common destination by far was Canada, followed by Great Britain and then Belgium. Very few CEOs have lived and worked in Latin America or the Far East.

If a firm must compete with companies in Japan, South Korea, Germany, or Taiwan, the best way to gain a comprehensive view of how they operate is to send the firm's best and brightest to those countries. It is unsettling to think that of the top 1,000 firms in the United States, few have CEOs who have worked abroad. This is even more unsettling when you consider recent research (Carpenter et al., 2001) indicates that CEOs with 6 or more years of international experience deliver significantly more value in firm performance and shareholder value than CEOs with little or no international experience. This same research found that there were additive effects on firm performance when the top five executives in the company also had experience working in another country.

If executives are to formulate valid global strategies for their firms—both at headquarters and at subsidiaries—they need to have an international perspective. Otherwise, a "garbage in, garbage out" phenomenon occurs in strategy formulation. Strategies are only as valid as the ideas, concepts, and knowledge of those formulating these strategies. Most executives would agree that the implementation of a global strategy is crucial to the success of a multinational or international company. Strategies do not implement themselves, however. People implement strategies.

To carry out a firm's worldwide strategic aims the way top management wants them to be carried out, the right people must be in place throughout the world. However, the right people for global assignments are not easy to find, and they do not automatically emerge out of the top 20 MBA programs. Instead, internationally astute people need to be developed. Ensuring that executives are willing to accept global assignments is simply too important to the future financial health of a firm to relegate developing global managers to a low priority on management's "worry list." As mentioned in chapter 1 (this volume), managers on global assignments play important strategic roles for MNCs: They coordinate between subsidiaries and headquarters, implement strategies, ensure the quality and effectiveness of organizational control systems, manage global information systems, and gain international and cross-cultural business skills that are critical to ensuring that top executive positions are filled by those with the necessary international experience and perspectives.

Figure 11.1 summarizes the key differences between taking a strategic and a tactical approach to international assignments. Interestingly, when executives we have talked with describe their approach in words that are more similar to those in the tactical-reactive column in Fig. 11.1, they tell us that none of them set out to take a tactical-reactive approach. This underscores the power of short-term pressures to divert recognition of the need to pursue a strategic and systematic approach to international assignments.

Strategic-Systematic	Tactical-Reactive
1. Approach international assignments as long-term investments.	1. Approach international assignments as short-term expenses.
2. Develop future executives with essential global perspectives and experiences to formulate and implement competitive strategies.	2. Focus on a quick-fix approach to a short-term problem in a foreign operation.
3. Increase the effectiveness of critical coordination and control functions between and among the home office and foreign operations.	3. Randomly and haphazardly perform some functions of international assignments and focus attention as problems arise.
4. Effectively disseminate information, technology, and corporate values throughout the worldwide organization.	4. Fail to systematically integrate worldwide organization in terms of values, technology, products, or brand.

FIG. 11.1. Contrasting approaches to international assignments.

Support Services Offered	Importance of Service	Satisfaction with Service	Satisfaction GAP
Cultural Orientation	72%	52%	20%
Language Training	57%	69%	12%
Homefinding/Settling-in	93%	64%	29%
Goods/Services Differentials	91%	73%	18%
Health Care	97%	65%	32%
Home Leave	92%	78%	14%
Ongoing Assistance Abroad	80%	53%	27%
Repatriation	91%	66%	25%
Spouse Counseling	71%	29%	42%
Career Planning	86%	27%	59%

FIG. 11.2. Satisfaction GAPS. Note: Results are based on a 1997 survey conducted by SHRM Institute for International Human Resources.

As Fig. 11.2 indicates, there is still a wide gap between the importance managers on global assignments place on various aspects of support a company could provide and these managers' satisfaction with the support they receive. In the following sections, we highlight some companies that are employing a more systematic and strategic approach to international assignments and reaping the rewards.

SELECTING: SELECTING THE RIGHT PEOPLE

Our first example of excellence comes from a human-relief organization, Worldvision International. This organization understands that its productivity is tied directly to the quality of the people it sends overseas to accomplish strategic objectives on short- and long-term assignments. Worldvision's philosophy is that the organization cannot afford to send the wrong people; thus, it spends significant time and energy on finding people who have the technical and cross-cultural attributes necessary for success in an international setting. Fewer than 50% of companies have structured selection systems, and less than 10% use any form of testing to screen candidates (Boles, 1997).

Worldvision's staffing policies, like those of any firm, are in a constant state of change and evolution, yet the one principle that remains constant is that effective employees are the key to overseas success. Worldvision International's selection process is divided into nine stages, which are summarized in Fig. 11.3.

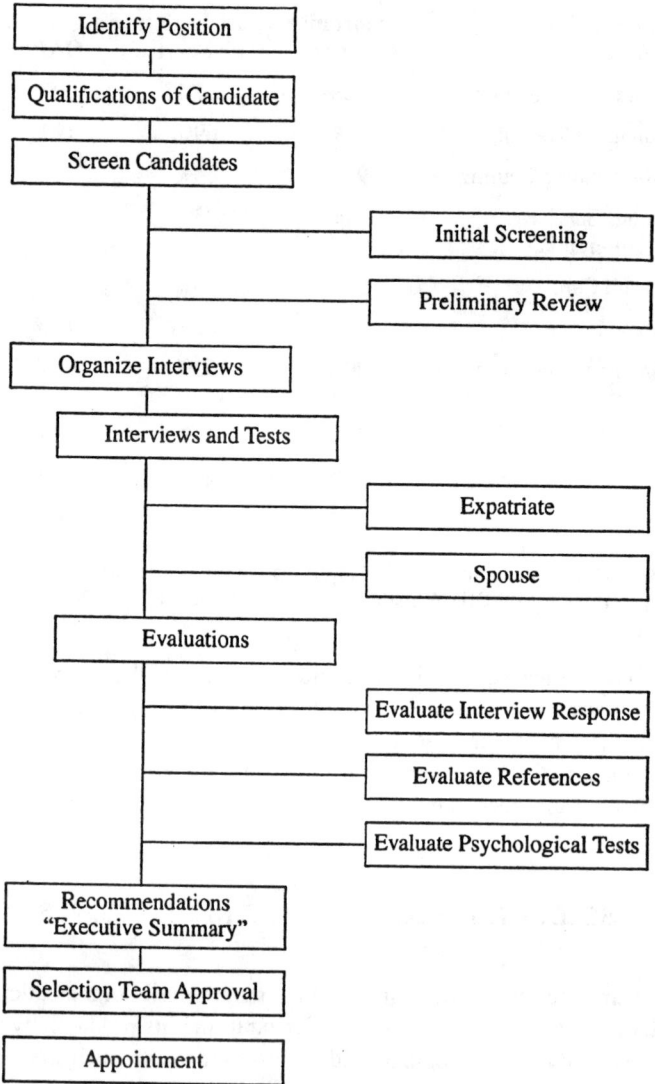

FIG. 11.3. The hiring process at Worldvision International.

Stage 1: Identifying the Position

In the first stage, the personnel department works with the hiring manager to clarify the nature of the position. Staff develop job descriptions and submit a personnel requisition to the division vice president, hiring manager, human resources director, and personnel director. All these individuals must agree on the nature and objectives of the international assignment before the search process begins.

Stage 2: Determining Candidates' Qualifications

In conjunction with the hiring manager, the personnel department determines the technical and cross-cultural qualifications required of candidates for the job. These qualifications include educational record, previous overseas background and experience, personality characteristics, and other specific attributes required to perform effectively in the position. Behavioral dimensions are identified, and a list of qualifications is developed that is used to assess candidates in interviews.

Stage 3: Screening

Using the criteria established in Stage 2, the personnel department conducts an initial screening of applicants. The best candidate is selected and presented to the hiring manager and the division vice president for a preliminary review. After the preliminary review, if the decision is positive—that is, everyone favors the candidate—the process moves to the next stage. If the preliminary review is negative, the process of screening candidates resumes.

Stage 4: Scheduling of Interviews

The staff in personnel work with the hiring manager to determine who should interview the candidate. Almost always, multiple interviews are scheduled. Interviews are generally conducted by staff from the personnel department, managers from the hiring department, and any other interviewers who have been selected.

Stage 5: Interviews and Tests

Candidates for global assignments can be evaluated on a variety of dimensions. From the following 28, interviewers select those dimensions most critical to the specific overseas assignment.

- Impact
- Oral presentation skills
- Management control
- Judgment
- Energy
- Listening skills
- Flexibility
- Perseverance
- Development of subordinates
- Initiative
- Decisiveness
- Motivation
- Resource utilization
- Financial analytical ability
- Communication skills—written
- Problem analysis
- Stress tolerance
- Negotiation
- Leadership
- Organizational sensitivity
- Technical translation
- Communication skills—verbal
- Delegation
- Adaptability
- Planning and organizing
- Sensitivity
- Political sensitivity
- Independence

A single interviewer will not cover all the selected dimensions; rather, each interviewer will be assigned to cover specific dimensions. Dimensions that are not deemed relevant to a particular assignment are ignored. Interviewers have specific questions to ask concerning each of their dimensions, although they may add other questions to the list.

Figures 11.4 and 11.5 include examples of questions interviewers may ask. When asking delegation-related questions, the interviewer must establish the purpose of the question—in other words, what the interviewee wanted to achieve and why the particular person was chosen to handle the matter.

Interviewers provide written evaluations of their interviews within 24 hr. In addition to written observations, interviewers are encouraged to offer two objective evaluations based on a scale ranging from 1 to 5: the degree to which each skill dimension was exhibited in the interview itself and an overall rating of the candidate as excellent, above average, average, below average, or poor.

1. What big obstacle did you have to overcome to get where you are today? How did you overcome it?
2. Have you ever submitted to your superior a good idea that he or she did not take action on? What did you do?
3. Can you relate an experience in which you believe you persisted too long? How could the situation have been improved?
4. Can you relate an experience in which you believe you gained something because you persisted for a length of time?
5. I see you did not complete —— course or activity. Why didn't you?
6. How long does it take you to complete an average project? What is the longest you have ever taken? Why?
7. How many times do you usually call on an official before you give up?
8. Describe a situation where you gave your all but failed.
9. What is the biggest project you did not successfully conclude? What did you do?
10. What university course gave you the most trouble? What did you do about it?
11. What was the biggest problem you encountered in college? On your last job? How did you handle the problems?

The interviewer should establish the size of problems or barriers in order to gauge the amount of perseverance required to overcome them.

FIG. 11.4. Perseverance.

1. Who is "minding the store" while you are here? How was that person selected? Why? How will you know how well he or she performed? Did you make a formal announcement to your subordinates concerning the responsibilities of the person you left in charge?
2. Explain your biggest mistake in delegating.
3. Explain your biggest mistake in not delegating.
4. What keeps you from delegating more?
5. Describe the type of decision making that you delegate to your subordinates.
6. Describe your criteria for delegating assignments.
7. If the degree of delegation varies among subordinates, explain how and why.
8. When did you have a major problem requiring staff help? What action did you take? Why did you ask particular people to assist you?
9. How much overtime do you put in per week? (Do you need to delegate more?)
10. After being away for several days, how do you familiarize yourself with the current situation in your organization?
11. Can you cite an example from your own experience where you were faced with delegating authority and/or responsibility? How did it work?

The interviewer should establish the purpose of each delegation mentioned. What did the interviewee want to achieve, and why was the particular person chosen to handle the matter?

FIG. 11.5. Delegation.

Stage 6: Evaluations of Interviews and Tests

The personnel department evaluates the written responses of the interviewers, the candidate's references, and the results of any psychological tests that have been administered during the interview process. The candidate's partner is also interviewed, although less intensively, and is given psychological tests to determine adaptability potential. The interviewer then establishes the size of any problems or barriers to gauge the amount of perseverance required to overcome them.

Stage 7: Personnel's Recommendations

The personnel staff summarizes the results of the interviews, references, and psychological tests and then presents a summary with recommendation to the hiring manager and the division vice president.

Stage 8: Approval

If the recommendation is to offer the position to the candidate, the next step is to have the candidate approved by the selection team consisting of the division vice president, hiring manager, human resource director, and personnel director. Personnel discusses the details of compensation and benefits with the hiring manager and then submits a personnel action form to the selection team for approval by each member of the team.

Stage 9: Appointment

The personnel department issues a letter of invitation that includes the job title, starting date, salary, terms of service, specific information related to the movement of household goods, banking arrangements, travel to the field assignment, and any other essential information. Although this process can be lengthy and time consuming, employees sent out under Worldvision's selection process have been rated as highly successful by the firm's top management. In Worldvision's experience, the investment of time and money in effective selection has paid off with significant long-term returns for the organization and the individual. Worldvision International directly applies many principles that were discussed in chapter 3 (this volume).

Worldvision initiated its selection process by involving a team of important "stakeholders," individuals concerned with the success of the organization's global managers. The organization then decided which strategic objectives a manager could accomplish overseas and developed a comprehensive, focused list of the selection criteria that were most relevant to a particular global assignment. Multiple selection methods were used to assess candidates against an array of selection criteria. Finally, Worldvision recognized the need to consider the families of candidates for global assignments and involved partners, as appropriate, in the selection process. The result of this approach is an impressive record of successful overseas assignments.

Essentially, Worldvision International uses a strategic-systematic approach to international staffing (versus a tactical-reactive approach). The differences between these approaches are illustrated in Fig. 11.6. Rather than being victimized by hasty selections based on a subsidiary's immediate need, Worldvision International carefully considers the nature and purpose of an assignment and then locates candidates who can effectively accomplish strategic purposes.

TRAINING: HELPING PEOPLE DO THE RIGHT THING

In chapter 4 (this volume), we pointed out that although predeparture training is important, in-country training—especially after 2 to 6 months—is critical to productivity while an employee is abroad. After the employee and partner have set-

Strategic-Systematic	Tactical-Reactive
Purpose	**Purpose**
• Executive development	• Fix a short-term problem
• Coordination and control	
• Transfer of innovation, technology, culture	
Focus	**Focus**
• Technical skills	• Technical skills
• Cross-cultural skills	
• Family situation	
Process	**Process**
• Multiple selection criteria	• Single selection criterion
• Multiple selection methods	• Single selection method

FIG. 11.6. Two approaches to selecting.

tled in, they are more motivated to try to understand the nuances of their new culture. At this time, rigorous training methods can produce long-lasting results. If employees are not exposed to the "meat" of the host culture during this period, they may develop inaccurate cognitive maps. As a result, rigid stereotypes or faulty attributional tendencies can become entrenched.

Once misunderstandings about the host culture are established, people are unlikely to actively seek out new and more accurate knowledge. To help firms avoid these problems, we recently offered in-country training to senior executives of IBM–Asia Pacific and IBM–Japan and their spouses who had been transferred to Tokyo. Our intensive, week-long seminar, which consisted of more than 60 hr of training, covered a variety of topics crucial to understanding the Japanese people and Japanese business culture including ancient, medieval, and modern history; Japanese religions and philosophy; socialization within Japan; psychological and sociological constructs of Japanese society; organizational behavior in the Japanese workplace; the business history of Japan; the organizational design and structure of Japanese corporations; the Japanese human resources system; business negotiations in Japan; strategies for learning the Japanese language; and relationships among business groups, the government, the education system, and the Japanese "mafia."

The three-step learning process described in chapter 4 (this volume) requires managers who are working abroad to first attend to the norms of the new culture, then retain and understand the nature of the new norms, and then practice new behaviors. By offering training in-country, IBM enhanced this learning process in that the executives were able to relate the concepts covered in the seminar to experiences they had just had the week or even the day before. They were also able to practice and validate what they were taught by implementing the knowledge immediately.

Consistent with the principles espoused in chapter 4 (this volume), IBM adopted the following instructional methods: lecture, films, books (homework assignments), cases, survival-level language training, role plays, and field trips. Each of these training methods was more effective in its delivery because the participants were studying Japan in Japan. Consistent with research findings and recommendations, spouses/domestic partners were included in this week-long training. IBM's approach to training international executives reflects its understanding of the need to invest in training to increase performance and adjustment as well as to increase the likelihood that the strategic objectives of the global assignment (e.g., executive development, coordination, corporate control, or technology transfer) will be accomplished. IBM's strategic-systematic approach to training is in direct contrast with the tactical-reactive approach outlined in Fig. 11.7.

Language Training

Language training is a thorny, complex issue for companies. Although many language firms claim that adults can learn a foreign language in 2 months by listening to tapes, the reality is that learning a foreign language requires dedicated effort. In fact, spoken fluency cannot generally be attained unless one lives overseas. Nevertheless, managers going on global assignments can acquire a minimal survival or conversational level before relocating. With that foundation, progress toward fluency can begin immediately. The danger of waiting to learn the language until arriving in a country is that fear, doubt, and work responsibilities can overwhelm the long, deliberate, day-to-day commitment learning a new

Strategic-Systematic	Tactical-Reactive
Investment Perspective	**Cost Perspective**
• Job training • Organization-business training • Culture training • Language training	• No post-arrival training
Results	**Results**
• Executive development • Improved current business results • Enhanced family adjustment	• Lower job performance • Higher expatriate turnover

FIG. 11.7. Two approaches to training.

language requires. With 4 to 6 months' notice before departing on a global assignment, it is quite possible, through classes or self-instruction, to reach the survival level in a language. Language instruction must occur for an hour or two daily before departure. How much can be accomplished depends, of course, on the ability of the learner to pick up languages, the techniques used to learn, and the difficulty of the language itself.

In our research and consulting, we have not come across any companies that offer a strategic, rigorous, daily language program for managers 4 to 6 months before departure—although language skills may be the key to accomplishing strategic objectives while on the assignment. One of the authors did put this principle of language preparation to the test on a personal basis, however. He received an assignment to spend a semester (approximately 4 months) in Switzerland. He studied German for an hour a day for 4 months. By departure date, he could speak better than at the survival level but not yet at the conversational level. After 5 months in Switzerland, he spoke at a conversational level. The ability to converse with the average person on the street increased his family's satisfaction with the overseas assignment, reduced stress, impressed the host nationals (who told him that the Americans they knew made no effort at all to learn German), and generally aided him in accomplishing his assignment there.

In-country, cross-cultural training plus language training conducted before departure and after arrival are indispensable to the effectiveness while working overseas. Each company must conduct training programs as best fit its needs, but the research to date shows that most companies are ignoring this very important aspect of international people management.

APPRAISING: DETERMINING WHETHER PEOPLE ARE DOING THINGS RIGHT

Effective performance appraisal is an important component of a successful global assignment. In chapter 7 (this volume), we discussed several factors that result in performance appraisal systems that benefit individuals and organizations. 3M has an exemplary appraisal process that incorporates evaluative and developmental components.

Primarily for evaluation purposes, 3M's performance appraisal system begins with input from the direct functional managers of all employees working overseas (who may or may not be host-country nationals). These managers work to define performance expectations for individuals and business units overseas on an annual basis. They also jointly define the set of performance expectations for each global manager's career development. After business and performance expectations have been defined consensually, these criteria are forwarded for review to the area vice president and the reentry (or repatriation) sponsor in the home country. Each of these individuals reviews the manager's strengths and points out potential problems in selecting performance criteria and career or individual developmental needs.

A similar "appraisal team" process is used to assess accomplishments against expectations. Initially, the accomplishments for the year are reviewed by the direct functional manager and the manager; the area vice president and the reentry sponsor also examine these assessments.

3M has also developed a parallel developmental performance appraisal process called the "Human Resource Review." On an annual basis, four individuals evaluate performance and career issues during global assignments. These individuals are the executive director of human resources for the international division, the area vice president, the area human resources manager or director, and a representative from the corporate executive resources department. During Human Resource Review meetings, these people assess expatriates along several dimensions. They examine performance during the past year, and more important, they consider major developmental issues. For example, a manager's career progression within the firm in terms of current and potential positions is often a point of discussion in the Human Resource Review meeting. This meeting also provides an established forum for managers to consider specific issues as they pertain to each overseas employee such as costs, assignment length, and repatriation.

3M's performance appraisal takes advantage of many principles discussed in chapter 7 (this volume). For example, in both the evaluative and developmental appraisal process, 3M involves more than one person. In fact, in assessing an employee's performance and career development, 3M incorporates multiple perspectives by including managers and executives who have a stake in the success of the employees they evaluate. Moreover, 3M not only examines performance-related issues, it also broadens the scope of the assessment to include career-related challenges global managers face. In other words, 3M's approach to appraisal contains three critical elements for success: (a) several individuals rate performance, (b) it incorporates a variety of assessment criteria, and (c) it focuses on both evaluative and developmental aspects of the appraisal process.

REWARDING: RECOGNIZING THE RIGHT THINGS PEOPLE DO

American Express has a well-conceived compensation program for its managers overseas, which is, in principle, an "equalization" approach. Specifically, American Express calculates housing and utility allowances by determining their costs in the country where the employee is to be assigned and explicitly considering the job, salary level, and family size of the employee. The housing allowance is provided in the currency of the host country, and the amount is adjusted for inflation annually or more frequently, depending on the rate of inflation in the host country.

Generally, American Express discourages employees from selling their houses in the home country or buying houses in the host country. However, the company does provide individuals with information as well as financial assistance for renting and managing their homes.

While the individual is being paid a housing and utility allowance in the host country, an estimate of the housing and utility costs the employee would have incurred in the absence of the global assignment is deducted from the individual's home-country salary. Consequently, individuals pay approximately the same amount for housing and utilities as they would have if they had not been overseas.

This approach also applies to taxes. An initial calculation of employees' home-country taxes is made at the beginning of the year, and a more detailed recalculation is made at the end of the year. This hypothetical tax is based on earned income, which includes salary and bonus. This earned income is reduced by any before-tax benefits and salary deferrals. Itemized deductions that individuals would probably claim if they were working in their home country are also calculated. This hypothetical tax is deducted from the individual's home-country salary over the course of the year. Adjustments deemed necessary because of the year-end recalculation are made at that time and are generally small. American Express then pays all home-country and host-country taxes actually incurred.

American Express also provides a number of other carefully considered allowances. In recognition of the difficulty of moving overseas and back again, American Express provides employees going abroad with a "mobility allowance." Part is paid when individuals go abroad and the rest when they return. The firm also pays for direct moving costs and provides a moderate sum to cover miscellaneous moving expenses. Additionally, American Express helps with educational expenses for children, home leave, and hardship allowances for selected countries that have extreme and difficult conditions.

This approach enables American Express to avoid rewarding A while hoping for B—that is, by following this approach, American Express is able to attract capable people to global assignments but does not make financial incentives the sole or major reason for considering or accepting an assignment overseas. Furthermore, paying host-country expenses in local currency while deducting expected home-country expenses reduces two potential problems. First, by paying expected costs in the local currency, the company reduces exchange-rate risks. Second, by deducting expected home-country costs, the company reduces the inequity that individuals might experience if all spendable and disposable income as well as allowance differentials were paid (and therefore visible) in the local currency. In addition to achieving these two objectives, these practices reduce the overall compensation and tax costs to the firm.

REPATRIATING: HELPING PEOPLE READJUST AND PERFORM

Less than one third of U.S. companies have a systematic repatriation program. The average turnover rate after repatriation in companies that have systematic programs is 5%, whereas it is roughly 25% in companies that do not have such programs. Given the lack of a strategic and systematic approach, it is not surprising that based on their most recent experience, only about 40% of international assignees would be willing to take another international assignment. In chapter 1 (this volume), we discussed GE Medical's "people" failure in a French subsidiary. Since that experience, GE Medical has worked hard to understand the international human resources process and in turn has created and implemented very effective programs for its managers overseas. The following section discusses some of the changes GE Medical has made in its repatriation process. Another company focusing increasing attention on repatriation is DaimlerChrysler. "From the Front Lines" addresses some of the changes the financial services arm has made recently.

One area that stands out is the attention GE Medical now pays to repatriates. For example, GE Medical initiated its Expatriate Sponsorship Program to ensure successful repatriation and to contribute to career development among managers who have done stints abroad. The program is straightforward: any manager who has completed a long-term global assignment (1 year or longer) is eligible for the program. He or she is then assigned a sponsor—a manager in the home country, preferably in the function in which the employee is most likely to return. There are four elements of the Expatriate Sponsorship Program: the repatriate, the sponsor, the host manager, and the Human Resource Network.

The Repatriate

Before going overseas, managers at GE Medical are expected to take responsibility for their own professional development. In other words, the company makes clear that it will not coddle employees throughout the assignment, and they will be required to expend effort to enhance their careers. Employees are expected to analyze the job opportunity in terms of its potential for personal career development, assess family readiness to live and work overseas, be clear about career expectations, prepare for the assignment by attending language and culture training programs with the family, and discuss future career opportunities with their sponsor on completion of the assignment.

While employees are overseas, they are expected to meet personally with their sponsor at least once a year. In this meeting, they are urged to discuss their own performance, career expectations, and other issues relevant to their situation in the firm. Managers who are working overseas are also expected to take advantage of company seminars, business meetings, reports, phone calls, and so on to build and maintain a network back at the home office. In addition, global managers are expected to work hard to enhance their management skills. Finally, before returning to the home office, global managers are expected to reassess their career options objectively, which is easier when a strong sponsor–expatriate relationship exists.

FROM THE FRONT LINES:

"Ensuring Higher Returns on Investment by Better Managing Repatriation"
by Shlomo Ben-Hur, Chief Learning Officer, and Kerstin Boecker, Director, Special Projects, DaimlerChrysler Services

At DaimlerChrysler Services, the financial services arm of Daimler-Chrysler, assigning managers to leadership and specialist positions around the world is a way of life. Traditionally, we have done a good job in identifying, placing, and integrating talented managers in other countries. However, from a survey conducted in 2000, we learned that we could clearly enhance our return on investment by focusing greater attention on the repatriation process. Subsequently, we discovered that what we really needed to do was plan for repatriation during the whole period when the manager was working abroad. Ultimately, improvements were made in the following areas:

- ❏ We increased the number of discussions with expatriates about their performance and potential. In addition to conducting discussions in the host country, we started conducting career-development discussions with expatriates in their home countries as well. Overnight, the number of discussions concerning performance, potential, and return dates increased twofold.
- ❏ We formed a development committee. Composed of all the heads of DaimlerChrysler's business divisions, this committee meets quarterly to discuss talent available in all parts of the business and its suitability for available jobs. Through this process, international assignees learn about far more options available to them at home. It has also opened more doors for employees to pursue assignments as third-country nationals.
- ❏ We identify mentors for international assignees in the home country. This ensures that managers working abroad are made aware of new developments at home and that the mentor is made aware of changes affecting the employee on the international assignment.

As part of our overall process of improving how we manage international assignments, we have started focusing on the repatriation process 9 to 12 months before the employee returns home. We have also expanded our definition of repatriates to include third-country nationals.

The Sponsor

The primary role of the sponsor is to ensure that the repatriate successfully integrates in the home country. Predeparture, the sponsor also has several responsibilities, including the following:

- Participating in the selection process.
- Studying the employee's repatriation plan to identify opportunities, roadblocks, and dead ends in the employee's career.
- Meeting with the expatriate to discuss the sponsor's views of the overseas assignment.
- Formally communicating the employee's repatriation plan to the Human Resource Network.

While the employee is overseas, the sponsor is required to monitor the employee's performance via phone conversations with the host manager and the global manager and by reviewing the employee's performance appraisals, to meet annually with the employee to discuss the assignment, to complete an evaluation of his or her performance for the Human Resource Network, and to encourage the employee to maintain a personal network at home. Before employees return from global assignments, sponsors are also expected to act as career advisers and to provide references to in-country managers on behalf of employees.

The Host Manager

The host manager plays an important role in the employee's life both during and after the global assignment. Before the manager arrives at his or her new home, the host manager assesses the impact of having an outsider in the workplace, plays a large role in the selection of the employee, prepares the organization to receive the employee, and assigns the employee an in-country mentor. During the employee's tenure abroad, the host manager is required to complete an annual performance appraisal and send copies of it to the employee's sponsor and to the Human Resource Network. Host managers must spend more time communicating with and providing feedback to managers working abroad than with other employees. Before the employee repatriates, host managers must activate the Human Resource Network to begin the repatriation process (this begins no later than 6 months before the term of the assignment is over). Host managers must also complete an appraisal of the manager's overall performance in the assignment and through the Human Resource Network, send it to the sponsor.

The Human Resource Network

The Human Resource Network at GE Medical Systems performs the following functions:

- Organizes a planning review twice a year.
- Coordinates candidate identification and selection.
- Manages sponsorship assignments.
- Manages repatriation plans sent by sponsors.
- Acts as the focal point for all global staffing and global management.
- Manages the contractual aspects of global assignments.
- Organizes training programs.

During the assignment, the network coordinates performance appraisals, salary planning, and reviews of the employee's performance. The network also monitors communications among all employees who are working overseas, sponsors, and host managers; and it manages the terms and conditions of assignments. No later than 6 months before repatriation, or at the request of the host manager, the network activates the repatriation process including planning for and finding the returning employee an appropriate position. Very few companies plan for repatriation in as organized a way as GE Medical does.

GE Medical understands that supporting employees on global assignments is an effective way to ensure the firm a global advantage in the future. GE recognizes its employees working abroad through several links it has established between them and the home country such as sponsors, frequent communication, a clear process for repatriation, and planning for the return assignment. In contrast, most companies, as reflected in Fig. 11.8, adopt an out-of-sight, out-of-mind approach that stems from a tactical-reactive philosophy and results in "orphaned" employees during and after global assignments.

SPECIAL CONSIDERATIONS IN THE POST-9/11 AGE

Although no company wants to put employees at risk, personal safety is a very real concern for any organization sending people abroad. More than ever, ensur-

Strategic-Systematic	**Tactical-Reactive**
Out of Sight ...With Support	**Out of Sight...Out of Mind**
• Organizational sponsor	• No support
• Communication links	• No planning of repatriation
• Visits home	
• Repatriation training and orientation	
• Systematic planning of return job	
• Proactive individual and organization assessment of return opportunities	

FIG. 11.8. Two approaches to support.

ing employees' personal welfare must be a high priority. For this reason, travel, whether for short-term or long-term assignments, has to be managed with the same attention reserved for other critical business operations. In this section, we touch on some of the key elements of a proactive approach to employee safety and security.

Crisis Management Team

No company wants to have to scramble frantically, weighing response after response, when a crisis has occurred and employees are in danger. To avoid this situation, it is critical that companies form crisis management teams and formulate guidelines for handling emergencies before a crisis occurs.

Employee Training and Education

One of the best ways companies can help to keep their employees safe while they are abroad is by providing them with up-to-date information about potential risks in the places to which they have relocated. This can be presented as part of other predeparture and postdeparture training, as Web-based training, and even as briefings by phone or in person. In addition to being told about how they can protect themselves from being victims of crime, terrorism, or other dangers, employees should be given training in how to respond in the event of an emergency. Emphasis should be placed on the company's ability to provide help 24/7, whatever the problem, and of course, on ensuring the personal safety of the employee and family.

Process for Providing Updates

Almost as important as training employees is having advisers who are trained to help employees avoid problems. These advisers should be travel coordinators/agents and security-support personnel. They should be vigilant to and notify employees of location-specific precautions as well as recommended practices.

Case 14

The importance of having a crisis management team in place as well as providing employees with safety-related training became glaringly evident to the Marriott Corporation when one of its hotels became the site of a terrorist bombing. The bombing occurred in Jakarta, Indonesia, at about 12:45 in the afternoon on August 6, 2003.

At the time of the explosion, Marriott had approximately 20 people on its crisis committee, including senior representatives of most of the company's functional areas such as Operations, HR, and Legal. According to the company's guidelines, the committee's first task was damage assessment. The committee quickly obtained crucial information from the general manager of the hotel in Jakarta.

Also high on the list of priorities was arranging a way for families of guests of the hotel to obtain information about their loved ones. Thus, within 90 min of the explosion, Marriott provided a guest list to a company called FEI, which, according to a prearranged agreement with Marriott, was prepared to handle inquiries from families during a crisis. Marriott also provided FEI with a list of its employees on the property. In addition, Marriott made grief counselors available to employees and their families.

Although Marriott lost no regular employees in the bombing, three contract employees were killed. Marriott is quick to note, however, that the loss of life was kept to a minimum in large part because the hotel had taken some of the best security measures possible (metal detectors, bag searches, and a security guard at an entrance to the hotel). The security guard was hired to examine cars as they approached the hotel. In this case, he saw the vehicle coming up the driveway and noted that it was weighted down. He became concerned and approached the vehicle, when he saw that the driver looked like a suicide bomber. Had the guard been less alert, the bomber would have gotten 10 ft closer to the front of the hotel and more than likely many more people would have been killed in the explosion.

In addition to the previous recommendations, companies are urged to heed the following guidelines:

- Provide safe havens for international assignees and their families.
- Be culturally aware when developing a crisis management plan. For example, in some cultures, providing grief counselors is not appropriate.
- Consider a variety of methods to maintain communication.
- Find out the best methods of communication in times of crisis and build in redundancy (have more than one backup).

GLOBAL ASSIGNMENTS: A KEY TO EXECUTIVE SUCCESS AND GLOBAL COMPETITIVENESS

Throughout this book, we have traced the rationale, logic, and evidence for taking a comprehensive, systematic, strategically oriented approach to international assignments. In addition to the economic arguments for establishing an integrated and well-designed international human resources system, there is another equally important consideration: a company's ethical responsibility to its employees. One way to understand this responsibility is to consider the preparations that go into a successful military campaign. Before going into battle, soldiers, pilots, and sailors are well trained in the use of weapons, battle tactics, and strategies of war. In fact, most have undergone numerous battle simulations or training at a very high cost. Events such as Operation Desert Storm provide clear evidence of the value of preparation and training. Why do military and intelligence organizations spend so much time preparing their people? The obvious answer is that without such training and preparation, performance in the field would suffer—lives would be lost. It is unethical to send people into situations in which their lives could be put at risk without carefully selecting those who will go and then training and supporting these people. This logic seems clear in war, yet few companies follow it when sending individuals into unfamiliar places on business. We have seen firsthand the pain, stress, grief, and other psychological challenges many employees experience in overseas assignments if they receive little support and training from their organizations. The following quote summarizes some of our thoughts on this matter:

> One must wonder if it is ethical to uproot an individual or a family, send them across the Pacific or Atlantic oceans, and expect them to make their way skillfully through an alien business and social culture on their own. Perhaps American executives reason that an extraordinary compensation package makes the exchange fair and ethical. Living and working overseas involves adjustments and stresses of a high magnitude. Placing individuals in such conditions without giving them the tools to manage them seems not only economically costly to the firm and personally costly to the individuals, but simply wrong.

Of course, sending people from Chicago to work in Japan is not the same as sending them off to war. Managers do not put their lives on the line in the same way soldiers do; however, managers working overseas often do put their careers, psychological health, marriages, children's education, and other significant aspects of their work and nonwork lives at risk. This reality raises a serious question: Is it ethical for a multinational firm to send the wrong people, to train them inadequately, or to fail to understand and support them while they live and work in another culture?

We feel that companies should offer sufficient support to employees throughout international assignments not only because this is the economically wise thing to do but, perhaps more important, because it is their ethical duty. Leading-edge firms take the whole business of globalization and internationalization seriously and invest heavily in internationalizing teams of senior executives and high-potential managers—the next generation of global leaders. These same companies also provide international assignees with the training, tools, and support to accomplish strategic objectives for their firms and to wage victorious campaigns against global competitors.

ESTABLISHING BEST PRACTICES

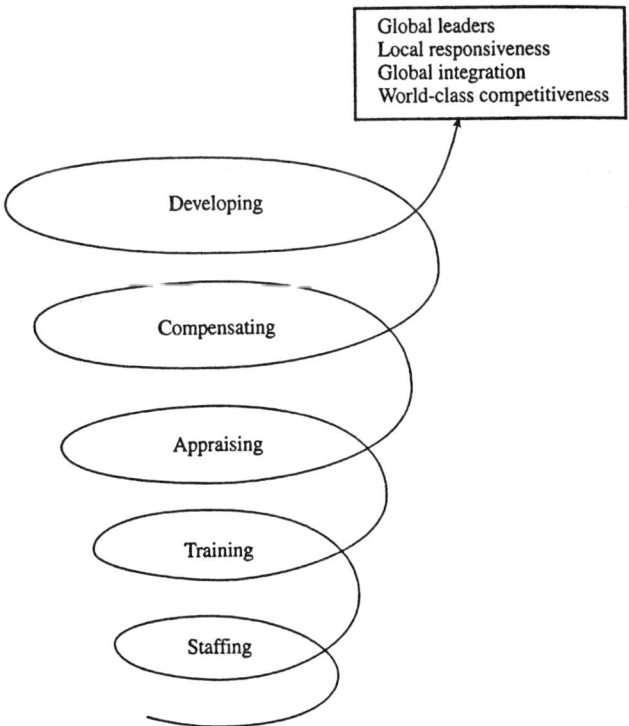

FIG. 11.9. Upward competitiveness spiral.

Multinational firms have strategic objectives that require employees to span the globe. To implement those objectives successfully, these firms rely on human as well as financial assets. Battle-hardened global companies have top executives and senior human resource managers who value and understand the need for a strategic-systematic approach to selecting, training, appraising, compensating, and developing people in international assignments. Figure 11.9 illustrates what happens when policies regarding global assignments reflect the attitude that these assignments are key to establishing and maintaining a competitive advantage and to developing an excellent executive team. The figure shows an upward spiral of global competitiveness. Institution of each policy increases the potential for success at the next stage.

In many ways, this spiral of policies and practices is similar to the board game Chutes and Ladders in which one wrong roll of the dice can mean a serious slide downward and loss of the game. In the global manager's world, one inappropriate policy can result in individual and organizational failure. By contrast, a strategic and systematic approach to global assignments helps ensure that managers working overseas do not slide down the policy "chutes" but climb up a "ladder," able to accomplish strategic objectives while overseas and after coming home. The result of a strategic-systematic approach is superior job performance overseas along with higher cross-cultural adjustment, higher levels of performance in foreign subsidiaries, successful adjustment and performance after repatriation, and utilization of international experience, skills, and expertise in the future. Successful international assignments through successful international human resources policies will be at the heart of successful international firms in the global marketplace of the 21st century.

References

Abe, H., & Wiseman, R. L. (1983). A cross-cultural confirmation of intercultural effectiveness. *International Journal of Intercultural Relations, 8,* 53–68.
Adler, N. J. (1981). Re-entry: Managing cross-cultural transitions. *Group and Organization Studies, 6,* 341–356.
Adler, N. J. (1987). Pacific basin managers: A gaijin, not a woman. *Human Resource Management, 26,* 159–192.
Adler, N. J. (2001). *International dimensions of organizational behavior* (4th ed.). Cincinnati, OH: South-Western College Publishing.
Adler, N. J. (2002). Global managers: No longer men alone. *International Journal of Human Resource Management, 13,* 743–760.
Adler, N. J., & Izraeli, D. N. (1988). *Management worldwide.* Armonk, NY: Sharpe.
Alampay, R. H., Beehr, T. A., & Christiansen, N. D. (in press). Antecedents and consequences of employees' adjustment to overseas assignments. *Applied Psychology: An International Review.*
Armstrong, M. (1994). *Performance management.* London: Kogan-Page.
Au, K. Y., & Fukuda, J. (2002). Boundary spanning behaviors of expatriates. *Journal of World Business, 37,* 285–296.
Aycan, Z. (1997). *Expatriate management: Theory and research.* Greenwich, CT: JAI.
Bandura, A. (1977). *Social learning theory.* Englewood Cliffs, NJ: Prentice Hall.
Beer, M. (1981). Performance appraisal: Dilemmas and possibilities. *Organizational Dynamics, 10,* 24–36.
Bell, N., & Staw, B. (1989). People as sculptors versus people as sculpture: The roles of personality and personal control in organizations. In M. B. Arthur, D. T. Hall, & B. Lawrence (Eds.), *Handbook of career theory.* Cambridge, England: Cambridge University Press.
Bird, A., Osland, J. S., Mendenhall, M., & Schneider, S. C. (1999). Adapting and adjusting to other cultures: What we know but don't always tell. *Journal of Management Inquiry, 8,* 152–165.
Björkman, I., & Gersten, M. (1992). Selecting and training Scandinavian expatriates: Determinants of corporate practice. *Scandinavian Journal of Management, 9,* 145–164.
Black, J. S. (1988). Work role transitions: A study of American expatriate managers in Japan. *Journal of International Business Studies, 19,* 277–294.

Black, J. S. (1989). Repatriation: A comparison of Japanese and American practices and results. In *Proceedings of the Eastern Academy of Management international conference* (Vol. 1). Hong Kong: Eastern Academy of Management.

Black, J. S. (1990). Factors related to the adjustment of Japanese expatriate managers in America. *Research in Personnel and Human Resource Management, 5,* 109–125.

Black, J. S. (1991, August). *A tale of three countries.* Paper presented at the annual meeting of the Academy of Management, Miami, FL.

Black, J. S. (1994). O Kaerinasai: Factors related to Japanese repatriation adjustment. *Human Relations, 47,* 1489–1508.

Black, J. S., & Gregersen, H. (1991a). The other half of the picture: Antecedents of spouse cross-cultural adjustment. *Journal of International Business Studies, 22,* 461–467.

Black, J. S., & Gregersen, H. (1991b). When Yankee comes home. Factors related to expatriate and spouse repatriation adjustment. *Journal of International Business Studies, 22,* 671–695.

Black, J. S., & Gregersen, H. (1992). *Functional and dysfunctional turnover after international assignments.* Unpublished manuscript, Amos Tuck School of Business Administration, Dartmouth College, Hanover, NH.

Black, J. S., & Gregersen, H. (1999). The right way to manage expats. *Harvard Business Review, 77*(2), 52–63.

Black, J. S., Gregersen, H., & Mendenhall, M. E. (1992). Toward a theoretical framework of repatriation adjustment. *Journal of International Business Studies, 23,* 737–760.

Black, J. S., & Mendenhall, M. E. (1989). Selecting cross-cultural training methods: A practical yet theory-based model. In M. Mendenhall & G. Oddou (Eds.). *Readings and cases in international human resource management.* Boston: PWS-Kent.

Black, J. S., & Mendenhall, M. E. (1990). A practical but theory-based framework for selecting cross-cultural training programs. *Human Resource Management, 28,* 511–539.

Black, J. S., Mendenhall, M., & Oddou, G. (1991). Toward a comprehensive model of international adjustment: An integration of multiple theoretical perspectives. *Academy of Management Review, 16,* 291–317.

Black, J. S., & Porter, L. W. (1991). Managerial behaviors and job performance: A successful manager in Los Angeles may not succeed in Hong Kong. *Journal of International Business Studies, 27,* 99–114.

Boles, M. (1997). How organized is your expatriate program? *Workforce, 76*(8), 21–22.

Bonache, J., Brewster, C., & Suutari, V. (2001). Expatriation: A developing research agenda. *Thunderbird International Business Review, 43,* 3–20.

Brett, J., & Stroh, L. K. (1995). Willingness to relocate internationally. *Human Resource Management Journal, 34,* 405–424.

Brewster, C. (1991). *The management of expatriates.* London: Kogan Page.

Caligiuri, P. (2000). The big five personality characteristics as predictors of expatriate's desire to terminate the assignment and supervisor-rated performance. *Personnel Psychology, 53,* 67–88.

Caligiuri, P. M., & Day, D. V. (2000). Effects of self-monitoring on technical, contextual, and assignment-specific performance: A study of cross-national work performance ratings. *Group and Organization Management, 25,* 154–174.

Caligiuri, P. M., Joshi, A., & Lazarova, M. (1999). Factors influencing the adjustment of women on global assignments. *International Journal of Human Resource Management, 10,* 163–179.

Caligiuri, P. M., & Tung, R. L. (1999). Male and female expatriates' success in masculine and feminine countries. *International Journal of Human Resource Management, 10,* 763–782.

Carpenter, M., Sanders, W., & Gregersen, H. (2000). International assignment experience at the top can make a bottom-line difference. *Human Resource Management, 39,* 277–285.

Carpenter, M., Sanders, W., & Gregersen, H. (2001). Bundling human capital with organizational context: The impact of international assignment experience on multinational firm performance and CEO pay. *Academy of Management Journal, 44,* 493–511.

Casio, W. F. (1986). *Managing human resources.* New York: McGraw-Hill.

Clarke, C., & Hammer, M. R. (1995). Predictors of Japanese and American managers' job success, personal adjustment, and intercultural interaction effectiveness. *Management International Review, 35,* 153–170.

Clegg, B., & Gray, S. J. (2002). Australian expatriates in Thailand: Some insights for expatriate management policies. *International Journal of Human Resource Management, 13,* 598–623.

Copeland, L., & Griggs, L. (1985). *Going international.* New York: New American Library.

Culpan, O., & Wright, G. (2002). Women abroad: Getting the best results from women managers. *International Journal of Human Resource Management, 13,* 784–801.

Deller, J. (1997). Expatriate selection: Possibilities and limitations of using personality scales. In Z. Aycan (Ed.), *New approaches to employee management: Theory and research* (Vol. 4, pp. 93–116). London: JAI.

Dennis, L. E., & Stroh, L. K. (1993). Take this job and . . . A case study of international adjustment. *International Journal of Organizational Analysis, 1,* 85–96.

Dowling, P., Schuler, R. S., & Welch, D. E. (1997). *International dimensions of human resource management* (3rd ed.). Belmont, CA: Wadsworth.

Downes, M., & Thomas, A. S. (1999). Managing overseas assignments to build organizational knowledge. *Human Resource Planning, 22*(4), 33–48.

Eschbach, D. M., Parker, G. E., & Stoeberl, P. A. (2001). American repatriate employees' retrospective assessments of the effects of cross-cultural training on their adaptation to international assignments. *International Journal of Human Resource Management, 12,* 270–287.

Feldman, D. C., & Bolino, M. C. (1999). The impact of on-site mentoring on expatriate socialization: A structural equation modeling approach. *International Journal of Human Resource Management, 10,* 54–71.

Gates, S. (1996). *Managing expatriates' return: A research report.* New York: Conference Board.

Gertsen, M. (1989). Expatriate training and selection. In R. Luostarinen (Ed.), *Proceedings of the European International Business Association conference.* Helsinki, Finland: European International Business Association.

Glisson, C., & Durrick, M. (1988). Predictors of job satisfaction and organizational commitment in human service organizations. *Administrative Science Quarterly, 33,* 61–81.

Goodman, N. (2003). *International joint ventures: Opportunities and enablers.* Retrieved: www.global-dynamics.com/GDIntJoint.html

Gregersen, H. B., & Black, J. S. (1992). Antecedents to commitment to a parent company and a foreign operation. *Academy of Management Journal, 35,* 65–90.

Gregersen, H. B., Black, J. S., & Hite, J. (1995). Expatriate performance appraisal: Principles, practices, and challenges. In J. Selmer (Ed.), *Expatriate management: New ideas for international business* (pp. 173–195). Westport, CT: Quorum Books.

Gregersen, H. B., Hite, J., & Black, J. S. (1996). Expatriate performance appraisal in U.S. multinational firms. *Journal of International Business Studies, 27,* 711–738.

Gregersen, H. B., Morrison, A., & Black, J. S. (1998). Developing leaders for the global frontier. *Sloan Management Review, 40,* 21–32.

Halcrow, A. (1999). Expats: The squandered resource. *Workforce, 78*(4), 42–46.

Hall, D. T., & Goodale, J. G. (1986). *Human resource management.* Glenview, IL: Scott, Foresman.

Hammer, M. R., Hart, W., & Rogan, R. (1998). Can you go home again? An analysis of the repatriation of corporate managers and spouses. *Management International Review, 38,* 67–86.

Harris, H., & Brewster, C. (1999). The coffee-machine system: How international selection really works. *International Journal of Human Resource Management, 10,* 488–500.

Harvey, M. (1998). Dual-career couples during international relocation: The trailing spouse. *International Journal of Human Resource Management, 9,* 309–331.

Harvey, M. R., Buckley, M. R., Novicevic, M. M., & Wiese, D. (1999). Mentoring dual-career expatriates: Sense-making and sense-giving social support process. *International Journal of Human Resource Management, 10,* 808–827.

Hoecklin, L. (1995). *Managing cultural differences: Strategies for competitive advantage.* Reading, MA: Addison-Wesley.
Hofstede, G. (1980). *Culture's consequences: International differences in work-related values.* Newbury Park, CA: Sage.
Ioannou, L. (1995). Unnatural selection. *International Business, 83*(8), 54–59.
Janssens, M. (1995). Intercultural interaction: A burden on international managers. *Journal of Organizational Behavior, 16,* 155–167.
Kainulainen, S. (1990). *Selection and training of personnel for foreign assignments.* Unpublished master's thesis, University of Vaasa, Vaasa, Finland.
Katz, J., & Seifer, D. M. (1996). It's a different world out there: Planning for expatriate success through selection, pre-departure training, and on-site socialization. *Human Resource Planning,* 32–47.
Kedia, B. L., & Mukherji, A. (1999). Global managers: Developing a mindset for global competitiveness. *Journal of World Business, 34,* 230–251.
Korn-Ferry International. (1981). *A study of the repatriation of the American international executive.* New York: Author.
Kraimer, M. L., Wayne, S. J., & Jaworski, R. A. (2001). Sources of support and expatriate performance: The mediating role of expatriate adjustment. *Personnel Psychology, 54,* 71–99.
Lazarova, M., & Caligiuri, P. (2001). Retaining expatriates: The role of organizational support practices. *Journal of World Business, 36,* 389–401.
Lindholm, N., Tahvanainen, M., & Björkman, I. (1999). Performance appraisal of host country employees: Western MNEs in China. In C. Brewster & H. Harris (Eds.), *International human resources management: Contemporary issues in Europe* (pp. 337–361). London: Routledge.
Linehan, M. (2000). *Senior female international managers: Why so few?* Aldershot, England: Ashgate.
Linehan, M., & Scullion, H. (2002). The repatriation of female international managers: An empirical study. *International Journal of Management, 23,* 649–658.
Linton, R. (1995). *The tree of culture.* Toronto, Ontario, Canada: McCleland & Stewart.
Lowe, K. B., Downes, M., & Kroeck, K. G. (1999). The impact of gender and location on the willingness to accept overseas assignments. *International Journal of Human Resource Management, 10,* 223–234.
Manz, C. C., & Sims, H. P. (1981). Vicarious learning: The influence of modeling on organizational behavior. *Academy of Management Review, 6,* 105–113.
Marx, E. (1996). *International human resource management in Britain and Germany.* London: Chameleon Press.
Mayrhofer, W., & Scullion, H. (2002). Female expatriates in international business: Empirical evidence from the German clothing industry. *International Journal of Human Resource Management, 13,* 815–836.
McGregor, D. (1960). *The human side of enterprise.* New York: McGraw-Hill.
Mendenhall, M., Dunbar, E., & Oddou, G. (1987). Expatriate selection, training, and career-pathing: A review and critique. *Human Resource Management, 26,* 331–345.
Mendenhall, M. E., Kuhlmann, T., & Stahl, G. (2001). *Developing global business leaders: Practices, policies, and innovations.* Westport, CT: Quorum Books.
Mendenhall, M., Kuhlmann, T., Stahl, G., & Osland, J. (2002). Employee development and expatriate assignments. In M. Gannon & K. Newman (Eds.), *Handbook of cross-cultural management* (pp. 155–183). Oxford, England: Blackwell.
Mendenhall, M., & Oddou, G. (1985). The dimensions of expatriate acculturation. A review. *Academy of Management Review, 10,* 39–47.
Mendenhall, M., & Oddou, G. (1986). Acculturation profiles of expatriate managers: Implications for cross-cultural training programs. *Columbia Journal of World Business, 21*(4), 73–79.
Mendenhall, M., & Stahl, G. K. (2000). Expatriate training and development: Where do we go from here? *Human Resource Management Journal, 39,* 251–265.

REFERENCES

Miller, E. (1973). The international selection decision: A study of managerial behavior in the selection decision process. *Academy of Management Journal, 16,* 239–252.

Mowday, R., Porter, L., & Steers, R. (1982). *Employee organization linkages: The psychology of commitment, absenteeism, and turnover.* San Diego: Academic.

Negandi, A. R., Eshghi, G. S., & Yuen, E. C. (1985). The managerial practices of Japanese subsidiaries overseas. *California Management Review, 4,* 93–105.

Nemetz, P. L., & Christensen, S. L. (1996). The challenge of cultural diversity: Harnessing diversity of views to understand multiculturalism. *Academy of Management Review, 21,* 434–462.

Nicholson, N., & Ayako, I. (1993). The adjustment of Japanese expatriates to living and working in Japan. *British Journal of Management, 4,* 93–105.

Oddou, G., & Mendenhall, M. (1991). Succession planning in the twenty-first century: How well are we grooming our future business leaders? *Business Horizons, 34*(1), 26–34.

Oddou, G., & Mendenhall, M. (2000). Expatriate performance appraisal: Problems and solutions. In M. Mendenhall & G. Oddou (Eds.), *Readings and cases in international human resource management* (pp. 213–223). Cincinnati, OH: South-Western College Publishing.

O'Hara, M., & Johansen, R. (1994). *Global work: Bridging distance, culture, and time.* San Francisco: Jossey-Bass.

Ones, D. S., & Viswesvaran, C. (1997). Personality determinants in the prediction of aspects of expatriate job success. In A. Zeynep (Ed.), *Expatriate management: Theory and research* (Vol. 4, pp. 63–92). London: JAI.

O'Reilly, C., & Chapman, J. (1983). Organizational commitment and psychological attachment: The effects of compliance, identification, and internalization of prosocial behavior. *Journal of Applied Psychology, 71,* 492–499.

Organization Resources Counselors. (1990). *1990 survey of international personnel and compensation practices.* New York: Author.

Paik, Y., Segaud, B., & Malinowski, C. (2002). How to improve repatriation management: Are motivations and expectations congruent between the company and expatriates? *International Journal of Management, 23,* 635–648.

Pellico, M. T., & Stroh, L. K. (1997). Spousal assistance programs: An integral component of the international assignment. In Z. Aycan (Ed.), *Expatriate management: Theory and research* (Vol. 4, pp. 227–244). London: JAI.

Porter, L., & McKibben, L. (1988). *Management education and development: Drift or thrust into the 21st century?* New York: McGraw-Hill.

Robinson, R. (1978). *International business management. A guide to decision making.* Chicago: Dryden.

Rodrigues, C. (1996). *International management.* St. Paul, MN: West.

Schein, E. (1984). Coming to a new awareness of organizational culture. *Sloan Management Review, 10,* 3–16.

Selmer, J. (1997). Differences in leadership behaviour between expatriate and local bosses as perceived by their host country national subordinates. *Leadership & Organization Development Journal, 18,* 13–22.

Selmer, J. (2000, August). *As time goes by: Does previous international experience facilitate expatriate adjustment?* Paper presented at the Academy of Management Meeting, Toronto, Ontario, Canada.

Selmer, J. (2001). Psychological barriers in adjustment and how they affect coping strategies: Western business expatriates in China. *International Journal of Human Resource Management, 12,* 151–165.

Selmer, J. (2002). Practice makes perfect? International experience and expatriate adjustment. *Management International Review, 42,* 71–88.

Selmer, J., Torbiörn, I., & de Leon, C. T. (1998). Sequential cross-cultural training for expatriate business managers: Pre-departure and post-arrival. *International Journal of Human Resource Management, 9,* 832–840.

Shaffer, M. A., Harrison, D. A., & Gilley, K. M. (1999). Dimensions, determinants, and differences in the expatriate adjustment process. *Journal of International Business Studies, 30,* 557–583.

Stahl, G. K., Miller, E. L., & Tung, R. L. (2002). Toward the boundaryless career: A closer look at the expatriate career concept and perceived implications of an international assignment. *Journal of World Business, 37*(3), 1–12.

Stening, B., Everett, J., & Longton, L. (1981). Mutual perception of managerial performance and style in multinational subsidiaries. *Journal of Occupational Psychology, 54,* 255–263.

Stroh, L. K. (1995). Predicting turnover among repatriates: Can organizations affect retention rates? *International Journal of Human Resource Management, 6,* 443–456.

Stroh, L. K. (1997, April). The family's role in international assignments (letter to the editor). *Fast Change Magazine.*

Stroh, L. K. (1999). A review of relocation: The impact on work and family. *Human Resource Management Review, 9,* 279–308.

Stroh, L. K., & Caligiuri, P. C. (1998). Increasing global competitiveness through effective people management. *Journal of World Business, 9,* 1–17.

Stroh, L. K., Dennis, L. E., & Cramer, T. C. (1994). Predictors of expatriate adjustment. *International Journal of Organizational Analysis, 2,* 176–192.

Stroh, L. K., Gregersen, H. B., & Black, J. S. (1998). Closing the gap: Expectations vs. realities among repatriates. *Journal of World Business, 33,* 111–124.

Stroh, L. K., Gregersen, H. B., & Black, J. S. (2000). Triumphs and tragedies: Expectations and commitments upon repatriation. *International Journal of Human Resource Management, 11,* 681–697.

Stroh, L. K., & Lautzenhiser, M. A. (1994). Benchmarking global human resources practices and procedures. *Mobility.*

Stroh, L. K., Northcraft, G., & Neale, M. (2002). *Organizational behavior: A management challenge.* Mahwah, NJ: Lawrence Erlbaum Associates.

Stroh, L. K., Varma, A., & Valy-Durbin, S. J. (2000). Why are women left at home: Are they unwilling to go on international assignments? *Journal of World Business, 35,* 241–255.

Suutari, V., & Brewster, C. (2001). Expatriate management practices and perceive relevance: Evidence from Finnish expatriates. *Personnel Review, 30,* 554–580.

Suutari, V., & Burch, D. (2001). The role of on-site training and support in expatriation: Existing and necessary host-company practices. *Career Development International, 6,* 298–311.

Suutari, V., & Tahvanainen, M. (2002). The antecedents of performance management among Finnish expatriates. *International Journal of Human Resource Management, 13,* 55–75.

Tahvanainen, M. (1998). *Expatriate performance management. The case of Nokia Telecommunications.* Helsinki, Finland: Helsinki School of Economics and Business Administration.

Takeuchi, R., Yun, S., & Russell, J. E. A. (2002). Antecedents and consequences of the perceived adjustment of Japanese expatriates in the USA. *International Journal of Human Resource Management, 13,* 1224–1244.

Thompson, R. W. (1998). Offer tax, financial advice. *HR Magazine, 43*(11), 10.

Torbiörn, I. (1982). *Living abroad: Personal adjustment and personnel policy in the overseas setting.* New York: Wiley.

Triandis, H. C., & Bhawuk, D. P. S. (1997). Culture theory and the meaning of relatedness. In P. C. Earley & M. Erez (Eds.), *New perspectives on international industrial/organizational psychology.* San Francisco: New Lexington Press.

Tung, R. L. (1981). Selection and training of personnel for overseas assignments. *Columbia Journal of World Business, 16,* 68–78.

Tung, R. L. (1982). Selection and training procedures of US, European, and Japanese multinationals. *California Management Review, 25,* 57–71.

Tung, R. L. (1988). Career issues in international assignments. *Academy of Management Executive, 2,* 241–244.

Vance, C. M., & Paik, Y. (2002). One size fits all in expatriate pre-departure training? Comparing host country voices of Mexican, Indonesian, and U.S. workers. *Journal of Management Development, 21*, 557–571.

Varma, A., & Stroh, L. K. (2002). A comparative study of the impact of leader-member exchange relationships in U.S. and India. *Cross-Cultural Management: An International Journal*.

Ward, C., & Kennedy, A. (1993). Where's the 'culture' in cross-cultural transition? *Journal of Cross-Cultural Psychology, 24*, 221–249.

Welch, D. (1994). International human resource management approaches and activities: A suggested framework. *Journal of Management, 31*, 139–164.

White, M. (1988). *The Japanese overseas.* New York: Free Press.

Windham International, National Foreign Trade Council, & Society for Human Resource Management. (2002). *Global relocation trends 2001 survey report.* New York: Windham International.

Woods, P. (2003). Performance management of Australian and Singaporean expatriates. *International Journal of Manpower, 24*, 517–534.

Zeira, Y., & Banai, M. (1985). Selection of managers for foreign posts. *International Studies of Management and Organization, 15*, 33–51.

Author Index

A

Abe, H., 57
Adler, N. J., 57, 61, 190, 201
Alampay, R. H., 111
Armstrong, M., 150
Au, K. Y., 117
Ayako, I., 53, 56
Aycan, Z., 103, 108, 110

B

Bahner, R., 30f
Banai, M., 54
Bandura, A., 82
Beehr, T. A., 111
Beer, M., 144–145
Bell, N., 99
Bhawuk, D. P. S., 35
Bird, A., 99
Björkman, I., 53, 54, 64, 147, 150
Black, J. S., 13–14, 15, 19, 50, 51, 52, 57, 58, 59, 60, 61, 62–63, 65, 82, 84, 85, 91, 110, 121, 124, 129, 139, 141, 147, 149, 150, 151, 155–156, 181, 189, 190, 191, 194, 198, 199, 200, 201, 203, 204, 205, 206, 212, 218, 219, 226, 235
Boles, M., 237
Bolino, M. C., 91, 93, 109, 112, 141

Bonache, J., 147
Brett, J., 68, 69
Brewster, C., 50, 51, 81, 147, 148–149, 150
Buckley, M. R., 109, 113
Burch, D., 81, 91

C

Caligiuri, P. C., 52, 58, 61, 67, 87, 100, 107, 193, 197
Caligiuri, P. M., 58, 59, 62, 102, 112, 147, 152, 218
Carpenter, M., 3, 27, 235
Casio, W. F., 144–145, 156
Chapman, J., 130
Christensen, S. L., 31
Christiansen, N. D., 111
Clarke, C., 57–58
Clegg, B., 100
Copeland, L., 17
Cramer, T. C., 65
Culpan, O., 58

D

Day, D. V., 147, 152
de Leon, C. T., 91, 93
Deller, J., 54

Dennis, L. E., 8, 37, 65
Dowling, P., 31
Downes, M., 107, 112
Dunbar, E., 84
Durrick, M., 130

E

Eschbach, D. M., 108
Eshghi, G. S., 57
Everett, J., 151, 156

F

Feldman, D. C., 91, 93, 109, 112, 141
Fukuda, J., 117

G

Gates, S., 66
Gersten, M., 53, 54, 64
Gilley, K. M., 100, 141
Glisson, C., 130
Goodale, J. G., 64
Goodman, N., 55
Gray, S. J., 100
Gregersen, H. B., 3, 13–14, 15, 19, 27, 50, 51, 52, 60, 61, 62–63, 65, 121, 147, 149, 150, 151, 155–156, 181, 189, 190, 191, 194, 198, 199, 203, 204, 206, 218, 219, 226, 235
Griggs, L., 17

H

Halcrow, A., 52, 199–200
Hall, D. T., 64
Hammer, M. R., 57–58, 201
Harris, H., 50, 51
Harrison, D. A., 100, 141
Hart, W., 201
Harvey, M. R., 109, 112, 113
Hite, J., 147, 149, 150, 151, 155–156, 181, 235
Hoecklin, L., 31
Hofstede, G., 31, 35

I

Ioannou, L., 54
Izraeli, D. N., 61

J

Janssens, M., 44
Jaworski, R. A., 109
Johansen, R., 110
Joshi, A., 102, 112

K

Kainulainen, S., 53
Katz, J., 53, 64
Kedia, B. L., 100
Kennedy, A., 46
Kraimer, M. L., 109
Kroeck, K. G., 112
Kuhlmann, T., 81

L

Lautzenhiser, M. A., 66
Lazarova, M., 102, 112, 218
Lindholm, N., 147, 150
Linehan, M., 60, 198
Linton, R., 31
Longton, L., 151, 156
Lowe, K. B., 112

M

Malinowski, C., 198, 214
Manz, C. C., 82
Marx, E., 81
Mayhofer, W., 109
McGregor, D., 33–35
McKibben, L., 235
Mendenhall, M., 60, 62–63, 81, 82, 84, 85, 91, 93, 94, 99, 149, 151, 152, 154, 194, 198
Miller, E., 52, 129, 219, 220
Mowday, R., 126, 130
Mukherji, A., 100

AUTHOR INDEX

N

Neale, M., 163
Negandi, A. R., 57
Nemetz, P. L., 31
Nicholson, N., 53, 56
Northcraft, G., 163
Novicevic, M. M., 109, 113

O

Oddou, G., 60, 81, 84, 85, 149, 151, 152, 154
O'Hara, M., 110
Ones, D. S., 54
O'Reilly, C., 130
Osland, J., 81, 99

P

Paik, Y., 92, 198, 214
Parker, G. E., 108
Pellico, M. T., 58, 72–73, 112, 113–114
Porter, L. W., 126, 129, 130, 235

R

Robinson, R., 149
Rodrigues, C., 190
Rogan, R., 201
Russell, J. E. A., 103

S

Sanders, W., 3, 27, 235
Schein, E., 33
Schneider, S. C., 99
Schuler, R. S., 31
Scullion, H., 109, 198
Segaud, B., 198, 214
Seifer, D. M., 53, 64
Selmer, J., 91, 93, 100, 102, 107, 108, 147, 151
Shaffer, M. A., 100, 141
Sims, H. P., 82
Stahl, G. K., 81, 91, 93, 94, 219, 220
Staw, B., 99
Steers, R., 126, 130

Stening, B., 151, 156
Stoeberi, P. A., 108
Stroh, L. K., 8, 15, 26, 37, 52, 57, 58, 60, 61, 62–63, 65, 66, 67, 68, 69, 70, 72–73, 87, 100, 107, 109, 112, 113–114, 163, 181, 191, 193, 194, 197, 226
Suutari, V., 81, 91, 147, 155

T

Tahvanainen, M., 147, 150, 155
Takeuchi, R., 103
Thomas, A. S., 107
Thompson, R. W., 69
Torbiörn, I., 86, 91, 93
Triandis, H. C., 35
Tung, R. L., 53, 56, 81, 84, 102, 219, 220

V

Valy-Durbin, S. J., 57, 60
Vance, C. M., 92
Varma, A., 26, 57, 60
Viswesvaran, C., 54

W

Ward, C., 46
Wayne, S. J., 109
Welch, D. E., 31
White, M., 53, 206
Wiese, D., 109, 113
Wiseman, R. L., 57
Woods, P., 147, 150, 155, 156
Wright, G., 58

Y

Yuen, E. C., 57
Yun, S., 103

Z

Zeira, Y., 54

Subject Index

A

Across-person equity, 161, 169
Adjustment process, *see* Cross-cultural adjustment; Repatriation
Africa, 86, 173, *see also specific country*
Airbus, 23
Alcoa, 21
Allegiances, *see* Organizational commitment
Allowances, *see* Compensation systems
American Express, 246–247
Anticipatory control, 100
Anticipatory expectations
 cross-cultural adjustment, 107, 108
 repatriation, 197, 202, 209
Appraisal, *see* Performance appraisal
Archer Daniels Midland (ADM), 77, 78
Artifacts, 29, 31, 37
Asia, 6, 36, 63, *see also specific country*
AT&T, 147, 148
Australia, 8, 9, 35, 85–86, 100, 195
Austria, 35, 209
Ausura, Maureen, 78
Avon, 3

B

Bahner, Richard R., 148
Baker & McKenzie, 162

Balance sheet compensation, 166–173
Behavioral reproduction, 83–84, 243–244
Belgium, 130–132, 231, 235
Ben-Hur, Shlomo, 249
bin Malek, Abd Mahmoud, 39
bin Mohamad, Mahathir, 40
Boecker, Kerstin, 249
Borneo, 38
Boyer, Earl, 82–84
Brazil, 49–51, 55, 62, 63
Brownouts, 13–14
Brunswick Corporation, 4

C

Canada, 57, 60, 174, 180, 198, 235
Car-driver allowance, 179
Career loops, 223–224
Careers
 compensation systems, 179–182
 dual-career marriages, 66–67, 68–69, 205–206
 free agents, 123–125
 mismanagement costs, 13, 16
 organizational commitment, 123–125, 131, 132, 133–134, 139–140, 143
 repatriation, 201, 248, 250
 retention, 223–224

SUBJECT INDEX

Careers *(cont.)*
  selection process
    candidate decision, 65–66
    dual-career marriages, 66–67, 68–69
Carlson, Gerald, 37–42, 45
Cendant International, 64
Chile, 149, 151, 153
China, 32, 72, 100, 209
  compensation systems, 172, 173
  organizational commitment, 122, 123
  performance appraisal, 153–154
  strategic roles, 6, 8, 15
Citicorp, 228, 229
Club membership allowance, 179
Coca-Cola Enterprises, 23, 71
Cognitive maps
  cross-cultural adjustment, 31–32, 37
  cross-cultural training, 83, 84, 243
  repatriation, 190, 191, 193, 197, 199
Colgate-Palmolive, 7, 71, 76, 228
Colombia, 35
Commitment, *see* Organizational commitment; Retention
Communication, *see* Information transfer
Communication skills, 57–58
Communication toughness, 85, 88–89, 90, 91
Compensation systems
  across-person equity, 161, 169
  allowances, 166, 175–179
    consequences, 179–180
  balance sheet approach, 166–173
    advantages, 172
    disadvantages, 172–173
    overview, 167*f*
  candidate decision, 66
  car-driver allowance, 179
  careers, 179–182
  children's education, 177, 247
  club membership allowance, 179
  composite market approach, 174
  cost-control objective, 161, 165–166, 175, 179–181, 184–185
    balance sheet approach, 172–173
    composite market approach, 174
    hardship allowance, 176
    parent-country equivalency approach, 173–174
    regional approach, 174
  cost-of-living allowances (COLAs), 170–172, 173, 174
  cross-cultural adjustment, 163–164, 184
  danger allowance, 176
  equity objective, 161, 164–165, 166, 181, 184–185, 246, 247
    balance sheet approach, 169–170, 172, 173
    composite market approach, 174
    parent-country equivalency approach, 173, 174
    regional approach, 174
  exchange rate, 172, 173, 181, 247
  expectancy theory, 182–184, 185
  foreign service premium (FSP), 175–176
  front-lines narrative, 162, 171
  hardship allowance, 176, 247
  home-leave allowance, 178, 247
  housing
    housing norm approach, 168–169
    job level approach, 169–170
  inflation rate, 171–172, 173, 181, 246
  local national approach, 174–175
  medical expenses, 178–179
  mobility objective, 161, 163–164, 166, 179–181, 182, 184–185, 247
    balance sheet approach, 166–168, 169–170, 172, 173
    composite market approach, 174
    danger allowance, 176
    foreign service premium (FSP), 175–176
    hardship allowance, 176
    home-leave allowance, 178
    local national approach, 175
    parent-country equivalency approach, 173
    regional approach, 174
  monetary incentives, 168, 175–176, 178, 179–180, 182, 214
  moving expenses, 176–177
  organizational policy, 181–185
  parent-country equivalency approach, 173–174
  people management, 18–19
    basic questions, 20*f*
  regional approach, 174
  relocation expenses, 177
  repatriation, 202–203, 214–215
  rest-relaxation allowance, 178
  retention, 225, 228
  rewarding A while hoping for B, 163, 164, 165, 166, 182, 184, 214, 247
  site allowance, 176
  standard of living, 166, 167*f*, 168–170, 172, 173–174, 175, 181
  strategic-systematic approach, 181–184, 185, 246–247

summary, 184–185
taxation
  equalization approach, 167–168, 247
  protection approach, 168
  vacation allowance, 178
  within-person equity, 161, 169
Composite market compensation, 174
Conflict-resolution skills, 57
Cost-control objective, *see* Compensation systems
Cost-of-living allowances (COLAs), 170–172, 173, 174
Crisis management team, 252–253
Cross-cultural adjustment
  anticipatory control, 100
  case study, 103–106
  compensation systems, 163–164, 184
  cultural assumptions, 31, 32, 99–100
    case study, 38–42
    environmental relationship, 33, 34*f*
    front-lines narrative, 36
    human activity, 34*f*, 35–37
    human nature, 33–35
    human relationships, 34*f*, 35, 36
    managerial implications, 34*f*
    reality/truth, 34*f*, 37
  cultural rules, 31–33, 99–100
  culture shock, 44
    violation of, 32–33, 44
  cultural values, 99–100
    artifacts, 29, 31, 37
    cognitive maps, 31–32, 37
    intangible aspects, 29, 31, 37
    matrix, 32*f*
    tangible aspects, 29, 31, 37
    tree analogy, 29, 30*f*, 31, 37
  culture shock, 44–46, 99
  dimensions of, 100–106
    overview, 101*f*
  dynamics of, 42–46
  influential factors, 106–114
    anticipatory expectations, 107, 108
    cross-cultural training, 107–108, 111, 113, 115*f*, 116
    cultural distance, 112–113
    cultural novelty, 103, 114
    foreign support, 112, 113, 114
    front-lines narrative, 109
    in-country, 107, 110–113
    individual, 107–108, 110
    international experience, 108
    job ambiguity, 111, 117
  job clarity, 111, 116–118
  job design, 115*f*, 116–118
  job discretion, 111, 116
  job-related, 110–111, 116–118
  logistical support, 111–112, 115*f*, 118
  mentoring programs, 109
  nonwork-related, 112–113, 114
  organizational, 107–109, 110, 111–112
  perceptual, 110
  predeparture, 107–109
  relational, 110
  role conflict, 111, 116–118
  self-oriented, 110
  spouse adjustment, 112, 113–114, 141–142
job environment, 100, 101*f*, 102
local environment, 100, 101*f*, 102
nonwork environment, 100, 101*f*, 103, 112–113, 114
organizational commitment, 130–131, 132, 141–142
predictive control, 99–100
repatriation, 190
retention, 226–227
routines, 43–44, 45
selection process, 50–51, 52, 53–54, 78, 114, 115*f*, 116
  family considerations, 50, 51, 52, 53–54, 78
success factors, 114–118
summary, 46, 119
women, 102
Cross-cultural training
  case study, 80–81, 89–94
  cognitive maps, 83, 84, 243
  cross-cultural adjustment, 107–108, 111, 113, 115*f*, 116
  framework for, 91–94
  in-country, 91, 93–94, 116, 242–244
  industry statistics, 81
  language, 244–245
  organizational commitment, 124, 132–133, 140, 141–142, 143
  organizational fit
    external fit, 95–96
    internal fit, 94–95
  organizational globalization patterns, 95–96
  people management, 18
    basic questions, 20*f*
  predeparture, 80–82, 91–92, 93, 116, 242

Cross-cultural training *(cont.)*
  program design
    communication toughness, 85, 88–89, 90, 91
    cultural toughness, 85–87, 90, 91
    family considerations, 91
    front-lines narrative, 87
    job toughness, 85, 89, 90–91
    principle application, 89–90
    program length, 84–85
    training rigor, 84, 85, 86, 88–89, 90, 91, 92, 93, 95f, 96, 116, 243
  safety, 252, 253
  selection process, 77
  strategic-systematic approach, 242–245
  tactical approach, 244f
  three-step learning process
    behavioral reproduction, 83–84, 243–244
    cultural attention, 82–83, 84, 243
    cultural retention, 83, 84, 243
  women, 87
Cultural assumptions, *see* Cross-cultural adjustment
Cultural attention, 82–83, 84, 243
Cultural distance, 112–113
Cultural novelty, 103, 114
Cultural retention, 83, 84, 243
Cultural rules, *see* Cross-cultural adjustment
Cultural toughness, 85–87, 90, 91
Cultural values, *see* Cross-cultural adjustment
Culture shock
  cross-cultural adjustment, 44–46, 99
  cultural mistakes, 44–45, 46
  cultural rules, 44
  defined, 44
  honeymoon phase, 44
  repatriation, 189–190, 191f, 194
  routines, 45
  self-image, 44–45, 46

D

DaimlerChrysler Services, 248, 249
Danger allowance, 176
Data Protection Act (Europe), 52–53
Denmark, 35
Dentsu Hogen K. K. Recol, 80–81, 89–94
Dorothy Syndrome, 193
Dow Jones & Company, 231
Downward-spiraling cycle

mismanagement costs, 16
organizational commitment, 125
Dual allegiance, *see* Organizational commitment
Dual-career marriages, 66–67, 68–69, 205–206
Dual citizens, *see* Organizational commitment
Duncan, Greg, 6
Duracell, 3
Dysfunctional retention, 219f, 221
Dysfunctional turnover, 219f, 220

E

Eastern Europe, 86
Eastman Chemical Company, 70–71
Economic Union, 6
Education, children's
  compensation systems, 177, 247
  selection process, 69
Employee Relocation Council (1997), 60–61
Equal Employment Opportunity Commission (United States), 52
Equity objective, *see* Compensation systems
Europe, 12, 86, 195, 216, *see also* Eastern Europe; *specific country*; Western Europe
  cross-cultural adjustment, 100, 102
  organizational commitment, 129–132, 223, 226–227
  repatriation, 189
  selection process, 52–53, 54, 56, 64
European Union, 6
Exchange rate, 172, 173, 181, 247
Expectancy theory, 182–184, 185
Export firms
  cross-cultural training, 95–96
  people management, 21

F

Failed assignments
  downtime costs, 13
  indirect costs, 13
  mismanagement costs, 12–13
  moving expenses, 12–13, 125
  organizational commitment, 124–125
  selection process, 50–51, 52
Family considerations, *see also* Repatriation; Selection process
  children's education, 69, 177, 247

SUBJECT INDEX

cross-cultural adjustment, 112, 113–114, 141–142
cross-cultural training, 91
dual-career marriages, 66–67, 68–69, 205–206
   organizational commitment, 141–142
   repatriation, 205–206
   safety, 69
   selection process, 66–67, 68–69
   spouse interviews, 53, 77
Fast Company Magazine, 109
Finland
   organizational commitment, 222, 223, 224, 225, 226, 227, 234
   repatriation, 190, 193, 194, 195, 198, 199, 201, 204, 205, 206, 208, 210, 211, 214, 215, 216
   selection process, 59, 61, 65, 66, 74
Flexibility, 59
Fluor Corporation, 7
Ford Motor Company, 9, 12, 139, 141–142, 212, 217
Foreign service premium (FSP), 175–176
Fortune 500, 4, 66
France
   compensation systems, 172
   organizational commitment, 125–126, 128, 129
   performance appraisal, 152–153, 157, 205
   selection process, 74–75
   strategic roles, 14–15, 16, 23
Fraser, Will, 8
Free agents, *see* Organizational commitment
Functional retention, 219*f*, 220
Functional turnover, 219*f*, 220

G

Gazzara, Kevin, 109
General Electric (GE), 6, 208
   Cie Générale de Radiologie (CGR) (France), 14–15, 16
   compensation systems, 165, 176
   Expatriate Sponsorship Program
      host manager, 250
      Human Resource Network, 250–251
      repatriate, 248
      sponsor, 250
   organizational commitment, 139–140, 141
   strategic-systematic approach, 248, 250–251

General Motors (GM), 3, 22, 64, 139, 141
Germany, 164, 173, 209, 218–219, 235
Gillette, 6–7
Global Assignment Preparedness Survey (G-A-P-S), 62–63, 77, 110
Globalization patterns, *see* Organizational globalization patterns
Global Relocation Trends Report, 15
Glynn, Amy, 231
Gone-native expatriates, *see* Organizational commitment
Great Britain, 35, 107, 209, 235
   compensation systems, 164, 173
   performance appraisal, 151–152, 153
   repatriation, 190, 195

H

Halo effect, 157, 158
Hardship allowance, 176, 247
Heart-at-home expatriates, *see* Organizational commitment
Hert, Ken, 231
High-risk repatriates, 208–209, 232
Home leave
   compensation systems, 178, 247
   repatriation, 198, 211
Honda Motors, 120–121, 139
Honeywell, 21
Hong Kong, 6, 72, 122, 129, 169
Host-country managers
   localized management, 11
   strategic roles, 11–12
Housing
   compensation systems
      housing norm, 168–169
      job level, 169–170
   repatriation, 204–205, 215
Howard, Marie, 171
Human Side of Enterprise, The (McGregor), 33–35

I

IBM, 22, 243–244
Implementor unit
   information transfer, 10, 136–137, 138
   organizational commitment, 136–137, 138
India, 6

Indonesia, 86, 122, 252–253
Inflation rate, 171–172, 173, 181, 246
Information transfer
   implementor unit, 10, 136–137, 138
   innovator unit, 10, 137, 138
   integrator unit, 10, 137, 138
   island unit, 9, 10f, 136, 137–138
   organizational commitment, 136–139
   repatriation, 197, 200, 209–211, 212, 213
   retention, 225–226, 227–229, 231
   safety, 252, 253
   selection process, 55
   strategic roles, 9–11, 136–139
Innovator unit
   factor endowments, 138
   information transfer, 10, 137, 138
   organizational commitment, 137, 138
Inpatriation, 22–23
Integrator unit
   information transfer, 10, 137, 138
   organizational commitment, 137, 138
Intel, 109
Internal fit, *see* Organizational fit
International commuting, 69–70
International experience
   cross-cultural adjustment, 108
   retention, 222–223, 225–226
   strategic-systematic approach, 235
International Hotel, 25
International joint ventures, 55
International Personnel Association (IPA), 60, 67, 74, 77
Interviews
   selection process, 53, 64, 77, 239–241
   spouse, 53, 77
Island unit
   information transfer, 9, 10f, 136, 137–138
   organizational commitment, 136, 137–138
Israel, 35

J

Japan
   compensation systems, 164, 170, 173
   cross-cultural adjustment, 37, 103–106, 107, 113, 116–117
   cross-cultural training, 80–81, 85–87, 89–94
   organizational commitment, 120–122, 128, 129, 132–134, 138, 139, 142, 220, 222, 223, 224, 225, 226–227, 233
   performance appraisal, 151–152, 153
   repatriation, 190, 195, 198, 199, 201, 202, 204, 205, 206–207, 208, 217
   selection process, 53–54, 56, 57, 61, 64, 65, 68, 69, 72
   strategic roles, 12, 13, 15
   strategic-systematic approach, 235, 243–244, 254
Job ambiguity, 111, 117
Job assignment
   repatriation, 199–200, 207–208, 211–212
   retention, 224, 228, 229–231
Job autonomy
   repatriation, 200–201, 211
   retention, 224, 229
Job clarity
   cross-cultural adjustment, 111, 116–118
   organizational commitment, 133, 142, 143
   repatriation, 200, 201, 207–208
   retention, 224, 229
   selection process, 238
Job demotion, 201
Job design, 115f, 116–118
Job discretion
   cross-cultural adjustment, 111, 116
   organizational commitment, 133, 142, 143
   repatriation, 201, 211–212
   retention, 224, 229
Job dissatisfaction, 189
Job offer, 77
Job promotion, 201
Job purpose, 76
Job-skill utilization
   repatriation, 202, 207–208, 211–212, 213
   retention, 219, 229–230
Job toughness, 85, 89, 90–91
Job turnover
   mismanagement costs, 15
   repatriation, 189, 191, 193, 194, 247–248
   retention, 218–219, 220

K

Keller, Randy, 169–170
KFC, 123
Kline & Associates, 103–106
Kodak, 11–12, 70–71
Kodak Australasia, 8
Koehler, Katie, 87
Korn-Ferry International, 151

L

Language
  cross-cultural training, 244–245
  repatriation, 194–195
Latin America, 6, 102, 235, *see also specific country*
Leadership development
  chief executive officers (CEOs), 3, 4
  front-lines narrative, 5
  organizational investment, 3–4
  selection process, 55
  strategic roles, 3–7
  succession planning, 3, 11
Leadership skills, 57
Li, Jenny, 127
L.L. Bean, 21
Local national compensation, 174–175
Logistical support, 111–112, 115*f*, 118
Lord Corporation, 74–75

M

Malay Dilemma, The (bin Mohamad), 40
Malaysia, 37–42, 45, 122
Managerial skills, 53, 56–57
Marriott Corporation, 87, 252–253
Mathison, 49–51, 55, 57, 62, 63
McDonald's, 43
Medical expenses, 178–179
Mentoring programs, 109
Mexico, 35, 36–37, 87, 102, 174
Microsoft, 33, 61
Middle East, 86, *see also specific country*
Mobility objective, *see* Compensation systems
Monetary incentives, 168, 175–176, 178, 179–180, 182, 214
Monsanto, 22, 71–72, 230
Motorola, 7, 72–73, 176, 213
Moving expenses
  compensation systems, 176–177, 247
  failed assignments, 12–13, 125
Multidomestic corporations (MDCs)
  cross-cultural training, 95–96
  executives, 22
  organizational commitment, 129, 138–139
  people management, 21–22
  retention, 232–233
  technical specialists, 22
Multifocal corporations, 23

Multinational corporations (MNCs)
  case study, 24–25
  cross-cultural training, 95–96
  organizational commitment, 134
  people management, 22–23, 24–25
  retention, 218, 219–221, 232–234
Murphy, John C., 213

N

Nairobi, 86
Neste Oy, 74
Nestle, 96
Netherlands, 120
New Zealand, 9, 83, 86
Neylon, John, 5
Nigeria, 95, 153
Nike, 8
Nokia, 128
North Korea, 8, 51, 61, 89, 116–117, 122

O

Organizational allegiance, *see* Organizational commitment
Organizational appreciation
  repatriation, 216
  retention, 225–226, 227–229
Organizational commitment, *see also* Retention
  allegiance patterns, 121–134
    advantages, 135–136*f*
    disadvantages, 135–136*f*
    management guidelines, 134–143
    overview, 122*f*
  dual citizens
    advantages, 133–134, 136*f*
    allegiance pattern, 132–134
    careers, 133–134, 143
    cross-cultural training, 132–133, 142, 143
    disadvantages, 136*f*
    information transfer, 137*f*, 138–139
    job clarity, 133, 142, 143
    job discretion, 133, 142, 143
    multidomestic corporations (MDCs), 138–139
    multinational corporations (MNCs), 134
    organizational policy, 134, 142–143
    role conflict, 133, 142, 143

Organizational commitment *(cont.)*
  free agents
    advantages, 123, 135*f*
    allegiance pattern, 121–125
    cross-cultural training, 124
    disadvantages, 123, 124–125, 135*f*
    downward-spiraling cycle, 125
    failed assignments, 124–125
    hired-guns, 122–123
    information transfer, 137*f*, 138
    plateaued-career, 123–125
    selection process, 124
  gone-native expatriates
    advantages, 126, 128–129, 135*f*
    allegiance pattern, 125–129
    careers, 139–140
    cross-cultural training, 140
    disadvantages, 126, 128, 129, 135*f*
    front-lines narrative, 127
    information transfer, 137*f*, 138
    multidomestic corporations (MDCs), 129
    organizational policy, 127, 139–141
    selection process, 139–140
    sponsorship, 126, 140
  heart-at-home expatriates
    advantages, 131–132, 135*f*
    allegiance pattern, 129–132
    careers, 131, 132
    cross-cultural adjustment, 130–131, 132, 141–142
    cross-cultural training, 141–142
    disadvantages, 131, 132, 135*f*
    family considerations, 141–142
    information transfer, 137*f*, 138
    organizational coordination/control, 131–132
    organizational policy, 131, 132, 141–142
    sponsorship, 131
  information transfer, 136–139
  strategic roles, 136–139
  subsidiary-expatriate match
    implementor unit, 136–137, 138
    innovator unit, 137, 138
    integrator unit, 137, 138
    island unit, 136, 137–138
Organizational coordination/control
  communication difficulties, 7
  cultural diversity, 8
  fragmentation, 8–9
  geographic dispersion, 8–9
  organizational commitment, 120–121, 131–132
  policy, 9
  resources, 9
  selection process, 55
  strategic roles, 7–9
  transportation difficulties, 7
Organizational ethics, 253–256
Organizational fit
  external fit
    cross-cultural training, 95–96
    defined, 24
    strategic roles, 24–25, 27
  internal fit
    cross-cultural training, 94–95
    defined, 24
    strategic roles, 24–25, 27
Organizational globalization patterns, *see also* Export firms; Multidomestic corporations (MDCs); Multifocal corporations; Multinational corporations (MNCs)
  cross-cultural training, 95–96
  people management, 20–25
  performance appraisal, 159–160
  retention, 219–221, 232–234
Organizational investment, *see also* Failed assignments
  leadership development, 3–4
  mismanagement costs, 12–17
  repatriation, 208, 211, 249
  retention, 218, 231
Organizational-needs analysis, 73, 74
Organizational policy
  compensation systems, 181–185
  organizational commitment, 127, 131, 132, 134, 139–143
  organizational coordination/control, 9
  people management, 22, 23, 26–27
  performance appraisal, 153–160
Organization Resources Counselors (ORC), 167

P

Pakistan, 35
Parent-country equivalency compensation, 173–174
Paterson, Rich, 5
PENNBANK, 37–42, 45
People management
  case study, 24–25
  dimensions of, 17–19, 21, 22, 23, 24, 26
  questions regarding, 19–20*f*

SUBJECT INDEX

framework for, 16–17
global context, 19–27
inpatriation, 22–23
key implications, 25–27
organizational globalization patterns, 20–25
  export firms, 21
  external fit, 24–25, 27
  internal fit, 24–25, 27
  multidomestic corporations (MDCs), 21–22
  multifocal corporations, 23
  multinational corporations (MNCs), 22–23, 24–25
organizational policy, 22, 23, 26–27
performance appraisal, 18, 159–160
Performance appraisal
  clients, 158, 159
  conflicts in, 145–146
  development goals, 145–146
  evaluation frequency, 158–159
  evaluation goals, 144–146
  halo effect, 157, 158
  invalid performance criteria
    organizational strategy, 153–155
    system design, 146, 147–150
  on-site managers, 157, 159
  organizational globalization patterns, 159–160
  organizational strategy, 153–160
  peer managers, 157–158, 159
  people management, 18, 159–160
    basic questions, 20f
  purpose of, 144–146
  rater bias, 147, 151–153, 158, 159
  rater competence
    evaluation wheel, 157–158
    organizational strategy, 155–158
    system design, 146, 150–151
    team approach, 156–158
  recency effect, 147, 159
  repatriation, 193–194, 217, 250, 251
  retention, 218, 219–221
  selection process, 50–51, 52
  self-appraisals, 158
  strategic-systematic approach, 159–160, 245–246
  subordinates, 158, 159
  summary, 160
  system design
    domestic context, 146–147
    front-lines narrative, 148
    international context, 147–153
  team approach, 156–158, 159, 160, 246

Perrin, Charles, 3
Peru, 88
Pfizer, 6
Philip Morris Companies, Inc., 6, 15
Philippines, 35
Pilarski, James, 87
Polynesia, 82–84
Predictive control, 99–100
PricewaterhouseCoopers (PWC), 5
Procter & Gamble, 21, 170, 171

Q

Quaker Oats, 70

R

R. Bahner International, 148
Recency effect, 147, 159
Regional compensation, 174
Reichert, Jack, 4
Relocation expenses, 177
Repatriation, *see also* Retention
  anticipatory expectations, 197, 202, 209
  careers, 201, 248, 250
  change components
    Dorothy Syndrome, 193
    home country, 191, 192f, 193, 194–195, 209
    organizational, 192f, 193, 194
    overview, 192f
    perceptual factors, 193, 209
    personal relationships, 193, 194, 204
  cognitive maps, 190, 191, 193, 197, 199
  cross-cultural adjustment, 190
  culture shock, 189–190, 191f, 194
  facilitation factors
    compensation systems, 202–203, 214–215
    home leave, 198, 211
    housing, 204–205, 215
    individual, 199
    information transfer, 197, 200, 209–211, 212, 213
    job-related, 199–202, 207–208, 211–214
    nonwork-related, 204–205
    organizational appreciation, 216
    overview, 196f
    perceptual, 199

Repatriation *(cont.)*
  facilitation factors *(cont.)*
    postreturn adjustment, 195, 196*f*, 199–205, 214–216
    prereturn adjustment, 195, 196*f*, 197–199, 211–214
    relational, 199
    repatriation team, 208, 211, 229, 230
    self-oriented, 199
    sponsorship, 197–198, 210, 248, 250–251
    supervisor empathy, 212, 214
    support groups, 216
    training, 198–199, 203, 210
    vacation time, 216
  family considerations
    dual-career marriages, 205–206
    family sponsors, 210
    information transfer, 210
    Japan, 206–207
    language, 194–195
    spillover effect, 194
    spouse adjustment, 207
  front-lines narrative, 213, 249
  high-risk repatriates, 208–209, 232
  job assignment, 199–200, 207–208, 211–212
  job autonomy, 200–201, 211
  job clarity, 200, 201, 207–208
  job demotion, 201
  job discretion, 201, 211–212
  job dissatisfaction, 189
  job promotion, 201
  job-skill utilization, 202, 207–208, 211–212, 213
  job turnover, 189, 191, 193, 194, 247–248
  organizational investment, 208, 211, 249
  organizational programs, 248, 250–251
  people management, 19
    basic questions, 20*f*
  performance appraisal, 193–194, 217, 250, 251
  strategic roles, 15, 16, 19, 22, 23
  strategic-systematic approach, 207–208, 247–251
    overview, 251*f*
  succession planning, 217
  tactical approach, 251*f*
Rest-relaxation allowance, 178
Retention
  cross-cultural adjustment, 226–227
  front-lines narrative, 231
  job assignment, 224, 228, 229–231
  job autonomy, 224, 229
  job clarity, 224, 229
  job discretion, 224, 229
  job-skill utilization, 219, 229–230
  job turnover, 218–219, 220
  multidomestic corporations (MDCs), 232–233
  multinational corporations (MNCs), 218, 219–221, 232–234
  organizational commitment influences
    career loops, 223–224
    compensation systems, 225, 228
    high-risk repatriates, 232
    individual factors, 221–224
    information transfer, 225–226, 227–229, 231
    international experience, 222–223, 225–226
    job factors, 219, 224, 228, 229–231
    nonwork factors, 226–227
    organizational appreciation, 225–226, 227–229
    overview, 222*f*
    parent-company tenure, 222
    sponsorship, 230
    training, 225, 228
  organizational globalization patterns, 219–221, 232–234
  organizational investment, 218, 231
  performance appraisal, 218, 219–221
  retention patterns
    dysfunctional retention, 219*f*, 221
    dysfunctional turnover, 219*f*, 220
    functional retention, 219*f*, 220
    functional turnover, 219*f*, 220
  role conflict, 224
  strategic assets, 218–219
  strategic-systematic approach, 227–232
Rewarding A while hoping for B compensation, 163, 164, 165, 166, 182, 184, 214, 247
Role conflict
  cross-cultural adjustment, 111, 116–118
  organizational commitment, 133, 142, 143
  retention, 224
Routines
  cross-cultural adjustment, 43–44, 45
  culture shock, 45
  disruption dimensions
    criticality, 44
    magnitude, 43–44
    scope, 43
  value of, 43
Russia, 5, 86

SUBJECT INDEX

S

Safety
  case study, 252–253
  family considerations, 69
  strategic-systematic approach, 251–253
    crisis management team, 252–253
    cross-cultural training, 252, 253
    guidelines, 253
    information transfer, 252, 253
Samsung, 7
Sara Lee Corporation, 70
Saudi Arabia, 88, 116–117, 167–168, 205
Scandinavia, 86, 214, *see also specific country*
selection process, 53–54, 56, 59, 61, 64, 65, 66, 74
Selection process
  candidate decision
    careers, 65–66
    compensation systems, 66
  candidate evaluation
    evaluators, 64, 240, 241
    family considerations, 64–65, 67
    selections tools, 61–62
    strategic-systematic approach, 240, 241
  case study, 49–51, 55, 57, 62, 63, 237–242
  common practices
    coffee-machine system, 51–52
    domestic context, 51–52
    international context, 52–54
  cross-cultural adjustment, 50–51, 52, 53–54, 78, 114, 115f, 116
  failed assignments, 50–51, 52
  family considerations
    candidate evaluation, 64–65, 67
    children's education, 69
    cross-cultural adjustment, 50, 51, 52, 53–54, 78
    dual-career marriages, 66–67, 68–69
    elder care, 69
    international commuting, 69–70
    lost income, 69
    relocation assistance programs, 69–73
    relocation willingness, 68, 69–70
    safety, 69
    spouse interviews, 53, 77
    systems perspective, 67–69
  international laws, 52–53
  narrowly-focused approach, 50–51
  organizational commitment, 124, 139–140
  people management, 17–18

  basic questions, 19f
  performance appraisal, 50–51, 52
  selection criteria
    categories of, 55–56
    communication skills, 57–58
    conflict-resolution skills, 57
    ethnocentricity, 59
    flexibility, 59
    leadership skills, 57
    managerial skills, 53, 56–57
    social skills, 58
    stability, 59–60
    strategic factors, 56, 76
    strategic-systematic approach, 76
    technical skills, 50–51, 52, 56–57
  selection tools
    biographical information, 62
    evaluation factors, 61–62, 77
    Global Assignment Preparedness Survey (G-A-P-S), 62–63, 77, 110
    interviews, 53, 64, 77, 239–241
    tests, 54, 62–63, 77, 239
    work samples, 63
  strategic-systematic approach
    candidate appointment, 242
    candidate evaluation, 240, 241
    candidate pool, 74, 76–77
    candidate pool development, 74–75
    candidate pool review, 76
    candidate qualifications, 239
    candidate screening, 239
    case study, 237–242
    cross-cultural training, 77
    cultural-context assessment, 76
    front-lines narrative, 78
    guidelines, 73–78
    information transfer, 55
    international context, 54
    international joint ventures, 55
    interviewer questions, 77, 240f, 241f
    interview schedule, 239
    job clarity, 238
    job offer, 77
    job purpose, 76
    leadership development, 55
    organizational coordination/control, 55
    organizational-needs analysis, 73, 74
    organizational success, 79
    overview, 238f, 243f
    personnel recommendations, 241
    selection criteria, 56, 76
    selection team approval, 77, 242

Selection process *(cont.)*
  strategic-systematic approach *(cont.)*
    selection team formation, 75–76
    specific-assignment candidates, 75–78
    sponsorship, 76
    testing dimensions, 77, 239–241
    value of, 54–55
  tactical approach, 77, 242, 243*f*
  women, 60–61
Singapore, 15, 72, 116–117, 122, 151–152, 176
Site allowance, 176
Social skills, 58
Sony Pictures Entertainment, 127
South Africa, 9, 108
South America, 3, 6, 86, *see also specific country*
Southeast Asia, 33, *see also specific country*
South Korea, 8, 51, 61, 89, 116–117, 122, 235
Sponsorship
  Expatriate Sponsorship Program (General Electric), 248, 250–251
  organizational commitment, 126, 131, 140
  repatriation, 197–198, 210, 248, 250–251
  retention, 230
  selection process, 76
Spouse, *see* Family considerations
Spyridakis, Maria, 172
Stability, 59–60
Standard of living, 166, 167*f*, 168–170, 172, 173–174, 175, 181
Stephens, Mel, 80–81, 89–94
Strategic roles
  host-country managers, 11–12
  information transfer, 9–11, 136–139
  leadership development, 3–7
  mismanagement costs
    brownouts, 13–14
    careers, 13, 16
    case study, 14–15
    downtime costs, 13
    downward-spiraling cycle, 16
    failed assignments, 12–13
    headquarter executives, 16
    indirect costs, 13
    job turnover, 15
    moving expenses, 12–13
    organizational investment, 12–14, 15, 16
  organizational commitment, 136–139
  organizational coordination/control, 7–9
  people management
    dimensions of, 17–19, 21, 22, 23, 24, 26

    framework for, 16–17
    global context, 19–27
    organizational globalization patterns, 20–25
    repatriation, 15, 16, 19, 22, 23
    strategic-systematic approach, 236
    summary, 27–28
    tactical approach, 4, 11
Strategic-systematic approach
  compensation systems, 181–184, 185, 246–247
  cross-cultural training, 242–245
  international experience, 235
  organizational ethics, 253–256
  organizational implementation, 235–236
  organizational services, 237
  performance appraisal, 159–160, 245–246
  repatriation, 207–208, 247–251
  retention, 227–232
  safety, 251–253
  selection process, 54–55, 56, 73–79, 237–242, 243*f*
  strategic roles, 236
  versus tactical-reactive, 236*f*
  upward competitiveness-spiral, 255–256
Succession planning
  leadership development, 3, 11
  repatriation, 217
Support groups, 216
Sweden, 86, 168, 214
Sweedlow, Doug, 49–51, 55, 62, 63
Switzerland, 169–170, 245

T

Tactical approach
  cross-cultural training, 244*f*
  repatriation, 251*f*
  selection process, 77, 242, 243*f*
  strategic roles, 4, 11
  versus strategic-systematic, 236*f*
Taiwan, 9, 122, 125, 235
Tatta, Raj, 5
Taxation
  equalization approach, 167–168, 247
  protection approach, 168
Teams
  performance appraisal, 156–158, 159, 160, 246
  repatriation, 208, 211, 229, 230

selection process
 candidate approval, 77, 242
 team formation, 75–76
Technical skills, 50–51, 52, 56–57
Technical specialists, 22
Tests, selection process
 selection tools, 54, 62–63, 77, 239
 strategic-systematic approach, 77, 239–241
Thailand, 100, 171–172
3M, 228, 245–246
Tompkins, Mike, 49–51, 55, 57, 62, 63
Training, *see also* Cross-cultural training
 repatriation, 198–199, 203, 210
 retention, 225, 228

U

United Kingdom, 72, 102, 201, 204, *see also* Great Britain
United States
 compensation systems, 164–165, 167, 169–170, 171, 172, 174, 176, 180
 cross-cultural adjustment, 102, 103–106, 113, 116–117
 cross-cultural training, 80–81, 86–87, 89–94
 cultural assumptions, 37
 cultural rules, 32
 human activity, 35–36
 internal/external fit, 25
 organizational commitment, 121–123, 124, 125–126, 128, 129–134, 138, 139–140, 141–142, 219, 220, 223, 224, 225, 226–227, 228, 229, 230, 231, 233
 performance appraisal, 148, 149, 151–154, 157
 personal power, 35

repatriation, 189, 190, 195, 197–198, 199–200, 201, 202, 203, 204, 205, 206, 208, 209, 212, 213, 214–215, 216, 217
 selection process, 52, 54, 56, 57, 59, 60, 61, 62, 64, 65, 68–69, 72, 74–75
 strategic-systematic approach, 235, 243–244, 245–251, 252–253, 254
Upward competitiveness-spiral, 255–256

V

Vacation
 compensation systems, 178
 repatriation, 216
Venezuela, 35, 197–198
Vietnam, 36–37, 190

W

Waggoner, Richard, 3
Weinger, Kerry, 162
Welch, Jack, 165
Western Europe, 8, 53, 54, 64, 86, 224
Within-person equity, 161, 169
Women
 cross-cultural adjustment, 102
 cross-cultural training, 87
 selection process, 60–61
Work samples, 63
Worldvision International, 237–242

Y

Yemen, 36–37